2021 Guide to Ultimate CYBERSECURITY

Written for Business Owners & Business Technology Managers

Author:

Josiv Krstinovski

Founder of KRSIT, a Managed Service Provider

COPYRIGHT

and to the benefit of the trademark owner, with no intention of infringement of the trademark. Where such designations appear in this book, they have been printed with initial caps.

The purpose of this book is to educate. The author and publisher do not warrant that the information contained in this book is fully complete and shall not be responsible for any errors or omissions. The author and publisher shall have neither liability nor responsibility to any person or entity with respect to any loss or damage caused or alleged to be caused directly or indirectly by this book.

DEDICATION

This book is dedicated to my mother and father who provided me with everything in life they were able to provide. My parents came to this country with very little, but they raised a great family. They gave us the ability to have a bright future for generations to come. They encouraged me to develop my talents and skills. I also want to dedicate this book to my older brother. He led me in pursuing a career in technology. He provided me the passion to follow into his footsteps.

ACKNOWLEDGMENTS

I want to express thanks the rest of my family and friends who inspired me and made me into the person I am today. Without the people in my life, family, mentors, professors, and colleagues this book would not be possible. I truly value the knowledge gained from these people. The experiences brought together from everyone who was a part of my adventures and journeys. Here is to all of you!

My brother, Tony Krstinovski, was a big influence in my pursuing a career in Information Technology. Our Aunt Zora Vlajkovic bought us our first computer in 1993. Together my brother and I learned a lot about computers at a young age and I have been following in the footsteps of my brother ever since.

My mom and dad provided, for both me and my brother, every bit of education they could afford so we could have a better life. My cousins, Billy and Christopher Kostovski have also been great friends and share my passion for technology throughout my youth and college years.

Other people who influenced me were by best friend, Chris Gaffney who was patient with me being on his computer all the time as a teen. Special thanks to Frank Vecsey, who played computer video games with me and sparked my passion for technology. Also special thanks to one of my best friends, Anthony Harwelik who shares my passion for technology and always has unique ways of incorporating technology in our lives.

I want to thank Best Buy for hiring me to have more experience practicing my skills on many of its customers to the customers being happy with the results I produced. My college professors at NJIT were very helpful in me achieving a Bachelor of Science.

My professional friend, Umut Bitlisli shares my passion and inspired me to write my own book. My great friend, Thomas Roman, is an entrepreneur who has inspired me to become more positive and think big. In turn, this friendship inspired him to do more with technology in his own business. I want to show my appreciation for James Kernan and Chris Wiser who developed masterminds events and training sessions that helped boost my business to the next level. I've learned from mentors such as Tony Robins, Darren Hardy, Robin Robins, Alex Rodgers from Chartec, and Paul Banco who gave me the opportunity manage CiBan technologies for him.

I appreciate Kate Frank for the hard work she did for this book. She was very valuable in helping pull this book together with her wordsmithing and editing skills.

Most of all, I would like to give special thanks to all of my team members who made KRS IT into what it is today: Saso Markoski, Kevin Obello, Darin Vaspol, Sean Sterling, Emmanuel Tejada, Gregory Coppola, Omar Hanafy, Mark Ilagan, Michael Cimilluca, Carlos Cruz, Chris Gaffney, and everyone else on the team.

ABOUT THE AUTHOR

My name is Josiv Krstinovski. As the founder and CEO of KRS IT Consulting, our focus has been, and always will be, focused on providing the highest quality IT services and solutions to our clients.

In fact, this book is an extension of the services our team provides every day. We believe the best clients are educated clients. If we can provide knowledge, on top of our record of exceptional service, our clients will stay with us and recommend our company to others. This respect for our clients is what has built KRSIT and it will continue to help us grow far into the future. KRS stands for Knowledge, Reliability, and Simplicity. These principles define our core professional and ethical values.

My Story

I started working with computers in the early 1990s. My brother was already working in IT, so it really helped me to follow in his footsteps. I either learned or inherited several techniques from him. First, I honed my skills by helping my family and friends during my teen years. Of course, it was mostly a labor of love. While I didn't charge to help them, I was able to grow my skills.

In 2004, I received my Bachelor of Science degree in Information Technology from the New Jersey Institute of Technology (NJIT.) My minor was in business management from the Rutgers/NJIT co-op program. While in school, I worked in several electronics stores such as Radio Shack and Best Buy. This time I was paid for my services.

After graduating, I held various IT consulting positions with companies like Future Generations and Alliant Technologies. This work led me to an opportunity to work for Vornado Realty Trust as a system engineer. I was part of the IT professional team for the help desk, desktop support, and network operations for locations in New York City, New Jersey, and Chicago.

In 2008, when the economy took a bad turn, I took the risk to start my own consulting firm, which has grown to become KRS IT Consulting.

The KRS IT Story

Through vast experiences in multiple categories, our team has become known as experts in the field of cybersecurity. My knowledge and experience go way beyond ordinary IT skills. Several firms depend on this expertise. I say this with utmost conviction. Honesty is my basic policy, and we all know it is also the best policy. Client satisfaction is always our top priority and we find our clients are delighted with the teams performance, year after year.

We focus on providing cutting-edge solutions to simplify a company's technology operations in a rapidly changing, technology-driven business environment. Staying up to date with the knowledge of the industry's leading products is part of our responsibility and we take it seriously. If we know what your business needs to stay competitive with technology, you have one less thing to worry about. It's why our clients love us.

Always Stay Safe!

Josiv Krstinovski

TABLE OF CONTENTS

Introduction

Information Technology (IT) has drastically changed the professional and business world of today. Luckily, most of the changes are for the better. However, there's a negative for ever positive in our world. Technology is no exception.

Risks to businesses are not new. But IT has just changed the shape of threats to business as it has changed the businesses themselves. So today, both businesses and individuals are up against cyber-threats.

In this book I'll detail these threats, along with their solutions. Cybercrime is increasing with the advancement of information technology. It is one of the primary reasons for me to choose this topic to share in the book. The book's core purpose is to provide effective solutions to all types and sizes of businesses.

Since cybersecurity threats come from all over the world every day, finding a reliable partner in cybersecurity, who knows how to keep you protected, should be the top priority of every business. In this book you will find updated and effective solutions to help you make your cybersecurity decisions wisely.

Cybercriminals are always looking for hacking opportunities. Millions of businesses face threats from these bad guys every year. They target businesses with weak security measures and hack into their computer systems with malicious intent.

For example, a criminal may hack into a system to steal information (data.) These activities are called hacking, phishing, ransomware, and many other names. All of them are bad. When this happens, it is not easy to recover. The criminal may sell the data or ask the victim business to pay a ransom. You will find more information about how and why the bad guys want your information later in the book. Just know it is never good for your business when cybercriminals can find access to your business data.

Breach of data reports come into KRS IT all the time. However, most companies are hesitant to report cyberattacks to authorities because it can be very damaging to their reputation. If they report a cybersecurity incident, it will be on the condition their identity will not be leaked, especially to the media.

Perhaps the worst damage of being a victim of a cyberattack is the loss of the client's trust.

Their caution is understandable because customers prefer firms with a positive image in the market. Firms who have been tarnished by a reported cybercriminal breach are not trusted. Every business knows their reputation matters a lot. If it is compromised, future business can be at stake. A bad reputation regarding cyberattacks is synonymous with scaring customers away and giving competitors a chance to take the lead. Perhaps the worst damage of being a victim of a cyberattack is the loss of the client's trust.

These days hacking business accounts has become very easy (and very profitable) for cybercriminals. In fact, the ones who are "experts in their field" can hack accounts without even triggering any suspicion. The companies with weaker security systems are the most vulnerable and are very easy to identify by the professional hacker.

The efforts of the pro hacker often result in illegal transactions and cases of fraud. These things are reported, at an ever-increasing level, every single year. This increase in activity clearly indicates the number of cybercrimes will likely increase in the future.

Top-level executives must realize the importance of cybersecurity. It's critical for keeping business data and accounts safe. While lower-level employees may not realize the sensitivity of this matter, this book is useful for all levels of employees and the executive decision-makers in the organization. You want everyone on the team to enhance their knowledge about cybersecurity so informed decisions can be made.

Here are some of the things executives and employees will learn from this book:

- The importance of cybersecurity
- Types of risk associated with cybersecurity
- Reasons why you need to protect against cybersecurity
- Consequences of a cyberattack
- Detailed solutions and prevention plans
- Which cybersecurity tools and software are more reliable and effective?

I'll expose many hidden realities and untold dark truths about the cyber-world in hopes of bringing you forward to a completely new perspective. My goal is to give you the knowledge and tools to address cybersecurity issues in an effective and timely way.

We've designed this book to be easy-to-understand so it can benefit all kinds of readers. The owners of small to mid-size companies (SMBs) will particularly benefit from this information. If you apply the suggested solutions in this book, it will strengthen cybersecurity for your business.

PART ONE

Cyberthreats & Cybersecurity

Overview

Cyberattacks are usually a catastrophe for your business. Below is a list of some of the ways it can wreak havoc on your company:

- Interruption of a company's production
- Closure of websites which can halt business activities
- Loss of vital data permanently
- Negative impact on brand reputation
- Often, the complete the failure of the business

As an IT professional, we witness frequent cyberattacks and we know how vulnerable companies and even governments can be in the face of these threats as the digital environment continues to advance. We cannot terminate cyberthreats for good. It's impractical to think we can. All the law-enforcing agencies in every country have not been able to slow down, much less cease this type of criminal activity.

In one way or the other, crimes take place. The law enforcing agencies act and try to take control of the criminals on a crime spree. They also keep an eye on known cybercriminals. They do prevent certain criminal events. Yet, a company's best defense is an IT partner who knows cybersecurity. When an event takes place, they are well-equipped with effective tools, and they know how to deal with cybercriminal activity.

Obviously, the cybersecurity equipment and expertise come at a price. In today's world, this price is as necessary as buying an insurance policy for the business. It needs to be positioned as a strategic area of running a business because of all the technological aspects of your organization.

Rules and Regulations

Business owners, executive boards, IT departments, and other decision-makers should realize legislators are attempting to change regulations to protect the public, including your company. While their efforts may be well meaning, it can also negatively impact your business. Whether it adds more costs to your operation, or protects people who don't have your company's best interest, you should be aware and assume responsibility for the increasingly strict cybersecurity rules and regulations being imposed on you and those who provide services to your company. You need to take this threat seriously.

Advanced Persistent Threat (APT)

First, let's define APT then talk about the major changes taking place in current cyberattacks. Wikipedia defines it as "a stealthy threat actor, typically a nation state or state-sponsored group, which gains unauthorized access to a computer network and remains undetected for an extended period."

While APT sounds bad enough, current hackers will make the deliberate decision to make themselves visible. It is the opposite of what happens with APT. These new, massive attacks aim to be particularly evident. This way, they expose the vulnerabilities of businesses and governments. It is the beginning of the threat which will become increasingly virulent and dangerous over time.

These frequent attacks alert businesses to take preventive measures to prevent such consequences. Vigilance should include their employees, partners, and the company's customers. As the line between personal and professional life becomes more and more entwined, companies should take responsibility for the digital life of their employees and their customers. It's important to realize how serious the situation has become and act. With intentional action, we can avoid these threats and create a secure digital space for the well-being of everyone.

CHAPTER ONE
What is Cybersecurity?

Security can be considered an absence of risks, having confidence in something, or being in a state of wellbeing. Academically speaking, cybersecurity is a discipline of Information Technology (IT) which studies causes and solutions to deal with threats and vulnerabilities of computer systems. This is especially true in networks where cyberthreats such as viruses, worms, trojan horses, identity theft, data theft, and phishing are common, along with other threats.

Cybersecurity Expert

A professional who detects cyberthreats and designs control/prevention systems against such attacks is a cybersecurity expert. They employ tools, processes, and strategies specific to the individual company who has hired them for this purpose. Here are some principles always considered:

- *The Integrity of the Data* – modification of any type of information must be known and authorized by the officially assigned person(s).

- *System Availability* – the continuous operation to maintain productivity and credibility of the business.

- *Confidentiality* – the disclosure of data must be authorized by the appropriate procedures and the data protected against attacks violating this principle.

In today's business environment, cybersecurity plays a very important role since all our data is in cyberspace – either in computers or digital networks. Precautions and measures to defend this data against potential attacks is the rallying call of our times. Remember, cybersecurity is not a fad. It is going to stay here. So, the need to protect the entire technological aspects of our lives has become essential to everyone since businesses and countries are dependent on technology.

Why Cybersecurity is Important

Today's business, to a greater or lesser extent, are immersed in the process of digital transformation. However, they face two contradictory realities. They must be online but being online carries a new risk – cyberattacks. To fight against this new breed of threats, every business must come up with a quicker and more powerful response.

Precautions and measures to defend this data against potential attacks is the rallying call of our times.

Having an online presence in various social networks has become a must for every business. Yet, this implies vital assets like reputation, finances, and the data of employees and customers are exposed to new risks. A programming error or a defect in a third-party software platform can open the door to leak millions of records. Even the giants of the business world like Equifax, Target, Home Depot, and social media king Facebook have been victims of such incidents in recent years. Attacks on the company data exposed the data of their millions of users.

Being competitive today requires the use of the latest tools such as cloud services and mobility technologies. IT departments have more heterogeneous and more complex systems to manage while their control over the systems is declining.

While is it easier to find security breaches, attackers have learned point the attacks toward the smaller suppliers who are less prepared rather than the mature companies with sophisticated security procedures. It will hurt the mature company indirectly in these cases.

This was the case of the data breach incident of Target in 2013 when the data of 40 million debit and credit card numbers was stolen. Target incurred losses of $200 million in a cyberattack with began with one of Target's suppliers.

In the cyber world, some risks are almost out of control for the average company:

- A hacker group can launch a DDOS (Distributed Denial-of-Service) attack and cause serious reputation damage, or loss of millions of dollars.
- A wrong comment, photo, or video on a social network can create a snowball and significantly damage a brand.
- A failure in a third-party software system (or a product such as Shellshock that left millions of servers vulnerable) can invalidate any defense.

Traditional Security

In the cyber-world, traditional security such as building a perimeter to defend your IT systems, creating firewalls on networks, and antivirus on PC's is necessary. But it is far from enough today.

In the IT security circle, there are two types of clients:

1. Those who know they have been victims and
2. Those who have been victims but still do not know.

With this being true, the focus much change from defenses to rapid detection and response. With the environment changing constantly, you need a professional partner who will share information and help you fight against cyberattacks.

Organized Cybercrime Industry

One step ahead of the defenses employed by business is an extremely advanced and organized industry. As the news of data breaches has flooded into our awareness in recent years, everyone (businesses and individuals) is vulnerable. It doesn't matter if you are a large or small company, a retail store, a bank, or a government, the risk is ever present.

Cybercriminals are organized and global. They use the latest means to collaborate, share, and achieve their malicious goals. However, if you have engaged true experts in cybersecurity, these providers are second to none in providing your best defense.

No business should work in isolation. The risks to one business can potentially affect many others. We discussed the case of Target earlier where one of their suppliers had a vulnerability which ended up costing the retail giant $200 million. Greater collaboration and information sharing will help achieve greater safety against such cyberthreats.

We all share the internet. This makes it easier for cybercriminals to gain access to our data illegally. Our networks and systems are therefore more vulnerable and less protected. Every company should recognize the value of the information it publishes online. It is essential to protect the data of the company, the customers, and the operation of the systems. To understand it more clearly, read the following points about why cybersecurity is extremely important and why you should invest in it.

- Preserve the data and protect the data against manipulation
- Protect the operability and integrity of IT systems
- Avoid illegal installation of spyware, malware, etc.
- Avoid trojan horses
- Avoid the activation of backdoors allowing hackers to take control of your systems
- Protect against loss/theft of personal, portable, and mobile devices

When the protection of your company's cybersecurity is left in the hands of qualified security professional, they can take care of securing your system optimally. It's the best way to stay one step ahead of hackers.

Types of Cybersecurity

Almost every company has networked computer systems with data stored in cyberspace. So, the protection of networked computers and online data is an obvious and important concern. You need to consider ever aspect of the risks involved. Since we need to connect to the internet, even when we don't have a website up and running or other forms of data, cyberthreats are looming around. We need to consider cybersecurity focusing on hardware, software, and network security. Let's look at these one at a time.

Hardware Security: Among all types of security efforts, hardware security offers the most robust protections. It can also help as an additional security filter to your most important systems. It involves both the physical protection and traffic control of a network. It can constantly monitor the whole system. The most typical examples of hardware security are firewalls, proxy servers, backups, battery banks for power outages, and cryptographic keys to encrypt, decrypt and authenticate systems.

Software Security: The biggest software security risk is the internet connection. This is true whether it is proprietary software or not. The entry doors for the cyberattacks is most often are implantation errors, design defects, and memory overflow. You want to protect the software from the first moment of its lifecycle or creation. This protection will help avoid security loopholes.

Software security is dedicated to blocking malicious hacker attacks. It is part of the process of implementing a program to prevent unauthorized modifications which can cause the malfunction or violation of the intellectual property of the program itself.

Network Security: This level of security protects the ease of use, reliability, integrity, and security of the network and data. It is designed to protect the network of computers and the devices connected to it. Protections from threats like viruses, worms, trojans, spyware, identify theft, data theft, zero-day attacks, hacker attacks, denial of service attacks and more are all part of this level of network security.

Network security is applied through the hardware and software of the system to accomplish various levels of security. This way, if the network is attacked, the rest of the security levels will continue to function. Components of this protection are antivirus, antispyware, and firewalls that cut unauthorized access. They include Virtual Private Networks (VPN), and Intrusion Prevention Systems (IPS).

Never forget the cyber attackers are highly skilled people. We have already witnessed data breaches of many businesses and government institutions. You need to secure every possible aspect of your IT systems with the help of the latest cybersecurity systems. To stay on top of this need, you need a professional cybersecurity expert.

CHAPTER TWO
What is a Cyber Attack?

A cyberattack consists of a series of illegal, criminal, unethical, or unauthorized actions. The objective is to destroy or compromise computer systems and data. A cyberattack can happen to individuals, organizations, and governments. Any computer system can be a resource to data and the criminals want illegal access and/or theft of the data.

The cyber attackers use electronic devices and the internet to violate, impair, or damage the data digitally. Their activities cause financial, physical, and intangible losses. These electronic crimes cover a wide spectrum of illegal actions and began before the legal systems even contemplated the scenario.

Legal instruments for controlling this type of crime have been gradually formulated and may need to be developed further. Cyber law is intended to protect the confidentiality of information and rights to property accessed over a computer network. People who commit computer crimes are experts in the use of technology. Their knowledge gives them the ability to not only know how to use them, but to be in work positions to facilitate their purposes. This means sometimes a company is cyberattacked by their own employees.

Cybercrimes are very difficult to identify. Yet fear of losing prestige is a valid reason for a company to not report they have been a victim of these crimes. Because of these facts, cybersecurity experts face difficulty understanding the scope of each crime and it's also why the global attacks are growing.

Each cyberattack can be carried out by a variety of actors:

1. ***Individuals,*** known as hackers, who act independently. They are generally motivated by financial gain.

2. ***Organized groups*** who are driven by other purposes including criminal terrorism, or ideological activism.

3. ***Governments*** engage in cyberwar strategies. They may target the computer network of other governments or important public/private assets.

4. ***Private companies*** who are hired for cyber surveillance activities. Many times, a corporation will pay a hacker to learn what a competitor is doing. It's profitable to get the information and use the data in their own strategies.

Primary Cyber Risks

Increasing dependence on the internet, along with new information/communication technologies, makes no one safe from a cyberattack.

- Vulnerabilities in the ***software*** used by companies and their employees. Every day, new software and security protocol flaws are discovered and exploited to carry out attacks. Although security service vendors react quickly by posting updates, the damage can be fatal.

- Bad ***configuration*** of information systems by companies, including servers, firewalls, and other systems.

- Bad security ***habits*** by employees, which allows cyber attackers to enter the information systems of companies for damaging or carrying out information theft, as in the case of ransomware.

- The use of ***personal mobile devices*** in corporate environments, known as BYOD or Bring Your Own Device! Though this trend brings great benefits to companies and their employees, it introduces significant risks if special security measures are not applied.

There are many ways to violate web applications to steal information from databases. Even if security measures are taken to resist those attacks. There are professionals in the business of generating new ways of cyberattacks. In addition, there are bad actors who respond to discontent with their superiors and will disconnect a network of banks, hospitals, or other organizations to create conflicts.

Risks to the systems are many, both internal and external. Even with no malicious intent, employees can get careless resulting in theft, sabotage, corruption, or violations of a computer system. Sometimes it can be blamed on insufficient training for using the system. It can also be human error or misuse of the access to information. When someone wants to do the same thing, but intentionally and from the outside, it is considered an external risk.

It can take a long time to detect an incident, especially when it is an internal breach of data. Better security practices to evaluate the suitability of data storage, strengthen access protection, and control access keys are essential to a good security practice.

Reasons Behind Cyber Attacks

The motivation for a cyber attack is as varied as the reasons why anyone would commit a crime. Each crime has a purpose which might be quite different than others, but they all have one thing in common...they have a use for the information technologies they hack. The hackers also vary in skill. Some may only be able to attack an IT system with fragile security. Others can hack into the most sophisticated security systems.

- ***Network Sabotage:*** This is a crime with a purpose to alter, modify, or delete information, programs, or files on the network. The result is a hindrance to the network normal operations. Common tools for this purpose are worms and malware. When these tools are used to destabilize a country, it is called cyberterrorism. This type of cyber attack generates a state of national shock with unquestionable ends.

- ***Espionage:*** When a company or government has data reserved as private, they can become the ideal target for this type of cybercriminal. It can be both a threat and a motive for the bad guys to gain unauthorized access to a network or system to obtain sensitive data. This damage or misuse of the data often refers to very important systems like economic, political, or military data.

- ***Fraud*:** Some hackers just want financial gain, so they obtain personal or bank data for illicit purposes. They may use unauthorized modification and impersonation to accomplish the crime. Sometimes they will use elements like Trojans or phishing. Other times it manipulates the input/output data for methods like salami slicing to divert a few cents from many accounts so there's no noticeable difference. A series of small, minor attacks will result in a larger attack when considered together.

- ***Unauthorized Access:*** Criminals may manage to enter protected electronic systems to modify or intercept files/processes. Wiretapping is an example of this sort of crime.

- ***Software Theft:*** Pirate activities compromise or undermine the rights of legitimate owners of software. When software has legal protection, but ends up being illegally distributed by criminals, many people are hurt in the process.

- ***Theft of Services:*** When a criminal gains access to your computer systems connected internet, it is easy for them to do illegal activity. These actions may be as simple as theft of time on your internet connection or could be the use of your digital services for illegal use of those resources. They might even impersonate someone who has been authorized as a trusted person for untrustworthy activities.

Consequences of Cyber Attacks

According to a **study** conducted by Kaspersky Lab and B2B International in 2017, cyberattacks cost an average of $1.3 million per company in North America. Since this number is 11% more than in 2016, it is fair to assume those costs have continued to rise in the past few years.

In the same year (2017) the average cost of a breach in North America was $1.3 million. One year later (2018,) the recovery cost had climbed up 23% to $1.6 million. These statistics were reported by **Kaspersky Lab** to the Canadian Underwriter online magazine.

The estimates cited above include both lost business cost, software and systems upgrades, expert advice, and extra expenses on internal staff. One thing is abundantly clear, costs and number of attacks is rising year over year. However, the asset most at risk is the business reputation. When a company does not know how to manage an attack, and the communication needed for customers and shareholders, the decline in reputation is even more costly.

In an Insights report published by **Forbes,** 46% of organizations had suffered reputational and brand value damage as a result of cyberattacks. In their conclusion they wrote: *Technology is the way that many clients, suppliers, and regulators interact with an organization. Any company that fails to* *live up to a promise of always-on service puts its reputation at risk. From underground hacker networks to complex supply chains, the threats to IT security and business continuity are proliferating—and so are the ways to deal with them. The best solutions extend beyond the boundaries of the IT department or business continuity team. A strong program for business resilience and IT security should work beyond traditional silos and extend across every part of an organization.*

In a PricewaterhouseCoopers report, they found: *"87% of consumers are willing to walk away and take their business elsewhere if, or when, a data breach occurs."* It would be no exaggeration to say a compromised reputation is synonymous with scaring your clients and customers away and giving your competitors a chance to overtake you.

If you were to ask me what is darker than a stain left by black paint, my answer would be a tarnished reputation is darker than the marks left by black paint. Paint on the walls could be scrubbed off, but a stain on a brand's image, left by a controversy, can never be erased. Every firm must be very conscious about their market image. When we established KRS IT by making a promise for **K**nowledge, **R**eliability, and **S**implicity, we knew we could never let this promise be tarnished. It is literally part of our company's name - and reputation. It's very important to us.

Every firm has to be very conscious about their market image and realize there are always people out there who are willing to destroy a reputation by exaggerating the rumors. The breach of data is harmful to the reputation of the firm because once such news is leaked trust is lost.

Unfortunately, most of the news related to breach of data are just rumors. But people who don't know it's simply an allegation end up relaying on those false stories and never investigate the matter in detail. Everyone is busy and such investigation is not important to them. They just believe the rumor and choose a competitor instead.

Cybercrimes are one of the major issues business owners' underestimate. They don't understand the sensitivity. The outcome can be the destruction of a brand image once the cybercriminals are successful with their mission.

People define threats to a business in different ways. I define threats as anything that slows down the progress of the business and distracts the business owners from the vision and mission of the firm. Any factor interrupting the business can be labeled a threat and must be addressed. Cyberattacks can cause a business to slow down, or in fact, cease their operations completely for a period of time.

As said earlier, threats can be categorized as internal, external, or both. They all need to be handled smartly and quickly. We see owners don't always consider internal threats as serious as those coming from the outside. However, this oversight can lead to a reduction in the firm's speed. The combination of financial losses and reputation damage is often fatal. According to an article in **Inc magazine** 60 percent of small businesses fold within six months of a cyberattack.

CHAPTER THREE
History of Cyber Breaches

In this chapter we will discuss some businesses, governments, and individuals who have been damaged via data breaches. We will also look at some of the individual hackers and groups. Evil almost always ends up with a bad outcome, sooner or later. Yet, in the world of internet hacking, some of the cybercriminals were given a chance to build their professional careers legally after they were convicted of a cybercrime. They started a "white hat hacking" practice for companies and governments. Let's look at who some of the more famous ones are, what they did, and what happened later.

Famous Hackers

Kevin Mitnick

In the late 1980s and early 1990s, Kevin Metnick earned the reputation of the worlds most famous hacker. He was the first to cause social alarm when he accessed the "ultra-protected" systems of Nokia and Motorola. He was on the FBI most wanted list before a well-publicized pursuit ended in an arrest at his apartment in 1995. In 1999, he pleaded guilty to a number of charges, was sentenced to prison for his crimes and now operates a very successful cyber consultant firm serving many of the Fortune 500 companies. His actions made permanent changes to the way companies protect their sensitive information.

Vladimir Levin

When banks started coming online, a Russian software engineer made breaking into banks his specialty. In 1995, Levin shocked authorities by re-routing $10 million dollars out of accounts at Citibank. He did this attack from his apartment in St. Petersburg, Russia. The money was transferred to multiple accounts around the world. He was arrested by Scotland Yard, sentenced to prison in the U.S and pay $240,000 to Citibank. It is assumed he got back to the pavilion of Russian hackers.

OurMine

Camouflaged after a supposed "security group", OurMine sells a service claiming they will scan the security of social media accounts and websites. Appearing to be "white hat" hackers, they vandalized accounts of celebrities' and organizations (NFL, Marvel, Google, Facebook, Netflix, Twitter, etc.) to advertise their services and to extort money from the victims.

Fancy Bears

This team of hackers is part of Russian intelligence and is known for getting involved in US presidential elections. They are also known as APT28, Tsar Team and other names. This cyber espionage group is thought to be responsible for leaking confidential medical files of Olympic athletes from the World Anti-Doping Agency (WADA) after several Russian athletes were excluded from the 2016 Olympics in Rio. These hackers are also well-known for link between them and the 2016 hacks of the Democratic National Committee and the Hilary Clinton campaign. Their methods of influence and disinformation include Zero-days (hacking computer developers,) spearphishing (targeted phishing to obtain something in particular by pretending to know the target person,) and malware.

The Shadow Brokers (TB)

In 2016, the Shadow Brokers become known by leaking tools and techniques from the US National Security Agency (NSA.) They publicly embarrassed the NSA and put important intelligence-gathering tools in the hands of many. Vulnerabilities in Microsoft Windows, Cisco routers, and Linux mail servers were all affected. They gave the authors of WannaCry ransomware, and Petya malware, the exploitation tools to infect hundreds of thousands of computers around the world.

Anonymous

An online movement called Anonymous is actually a diffuse, yet well-coordinated entity with a claim to defend "truth, freedom, and the end of censorship." They have attacked several governments, along with their institutions and agencies. Many people know them for their protests, pranks, and hacks targeting the Church of Scientology in the Chanology Project of 2008. Their antics are international, going wherever their loosely defined hacktivism leads them.

Lazarus Group

The U.S. FBI says this group is a sponsored by the North Korean government. In 2009, it became famous for the denial-of-service operations on U.S. and South Korean websites. While not much is known about the group, they are attributed responsibility for

- Sony hack in 2014 - confidential data from the film studio was leaked to public
- 2016 theft of $81 million from the central bank of Bangladesh
- Global WannaCry ransomware attack in 2017

Tarh Andishan

The Tehran secret service put a team of hackers into operation and in 2014, the team was responsible for Operation Cleaver. This compromised computers and networks from 16 different countries and targeted energy, health, logistics, defense contractors, airlines, universities, government agencies and more than 50 total organizations. It is thought this operation was a response to the Stuxnet worm which affected Iranian nuclear facilities. Any infrastructure in countries considered enemies of the Iranian state were at risk.

LulzSec

The malicious group of black hat hackers known as LulzSec gained visibility in May 2011. They coined the name from "lulz" for laughs, and "security" which is the target of the hacks. Their motto is "Laughing at your security

What I mean to show you here is being online exposes us to such sick minds. They destroy businesses and people... just for fun! Imagine how many more sick people could be out there.

since 2011." It is said none of the members met or even knew the identities of each other. Yet they hacked to express their ideology, be entertained, and also financial gain.

Their modus operandi was to access sensitive information of a company, like emails, user lists, and passwords. Then they would alter their public websites or publish confidential information on the internet. Various media credits LulzSec as being more as pranksters and hackers seeking to harm. Some mention this group for being triggered by the WikiLeaks case.

The American man who led this group before he was arrested, Hector Xavier "Sabu" Monsegur was sentenced to 124 years in prison. He served 7 months of the sentence and then given "time served" for cooperating with the federal investigations. This cooperation is credited for helping "disrupt or prevent at least 300 hacks" according to an FBI spokesperson.

LulzSec Hacking Victims

Fox.com: The first recorded attack by was Fox.com's website. In addition to leaking personal information of 73,000 contestants on X Factor. They also accessed the databases and leaked or altered the passwords and LinkedIn profiles of employees. They claimed their motivation as defending the rapper Common who had been allegedly called "vile" on the air.

Sony: A series of LulzSec attacks were launched against Sony Pictures. The private information of more than a million accounts of people's names, passwords, email addresses, home addresses, and birth dates. The also stole music codes and coupons to ensure maximum embarrassment to Sony. They said they used a SQL injection, one of the most common and primitive vulnerabilities. From a single injection, they claim to have accessed everything. They declared it would be the "beginning of the end for Sony". LulzSec claimed the data was extremely easy to find and exploit. Sony took legal action against George Hotz for jailbreaking (aka reverse engineering) the PS3. Later, the hacked data was used in scams.

Black & Berg Cybersecurity Consulting: This small network security firm challenged hackers to enter and alter their website. The successful hacker was to receive $10,000 reward if they accessed their website and positioned the senior advisor Joe Black. When LulzSec attacked the website, they left the message, "Done. That was easy. Keep your money. We do it for the Lulz."

PBS (Public Broadcasting System): In response to a Frontline story about Wikileaks, LulzSec said they were "less than impressed." So, they defaced the PBS.org website and published a fake story suggesting the murdered rapper Tupac Shakur was alive and living in New Zealand.

A Website with Adult Content: An X-rated porn site was hacked, and 26,000 user's login passwords were published publicly. They also exposed the credentials of other administrators of other adult websites.

InfraGard: The Atlanta chapter's database of Infragard, a private sector partner with the FBI, was hacked and the user database was uploaded to the internet. User email accounts were published and their website was altered.

Bethesda Game Studios: This game development studio was part of the Sony network. A list of 200,000 Brink accounts was obtained by the group as part of the Sony hack. They sent a message saying they were not publishing the stolen data because they liked the company. At the end of the message, they said "You're welcome."

CIA.gov: On June 15, 2011, the group took credit for an attack on the public website of the Central Intelligence Agency of America (CIA), an attack that made the website inaccessible to legitimate users for over two hours.

Gaming Sites: EVE Online, Minecraft, League of Legends, The Escapist, and FinFisher were taken down for a three-hour denial-of-service event on June 14, 2011.

SOCA: An agency in the UK by the name Serious Organised Crime Agency (SOCA) was attacked on June 20, 2011. This is considered the first of a group operation called

"Operation Anti-Security" where hackers were encouraged to steal and publish confidential information from government agencies. As proof of the group's involvement in the attack, they left the text "Anitsec". Some media called this event a "declaration of cyberwar" against the government and large corporations.

Brazilian Government: On June 22, 2011, it was announced that a possible Brazilian faction of the group attacked brazil.gov.br and presidencia.gov.br.

Arizona Public Safety: LulzSec published information they allege came from ADPS (Arizona Department of Public Safety.) They argued it was a form of political protest against illegal immigration in the state.

LulzSec Announces Dissolution

June 25, 2011, and under the pretext of meeting 50 days of attacks, the group LulzSec announced its dissolution. In a statement for their usual channels, amid much speculation in the media about what the real reasons were for why LulzSec dissolved. These are among the speculations:

- Government agencies hunted down the group's members
- Other groups of hackers hated LulzSec and they became a target of attacks themselves

- US reforms to double the sentences of perpetrators of cyberattacks. This reform was probably motivated by the actions of LulzSec during the 50 days

- Change of strategy, a way to regroup and reemerge under another name

These cyberattackers are relentless. They have no sympathy for any occasion. During the recent global COVID pandemic, they triggered ransomware called NetWalker (previously known as Mailto.) Hospitals were targeted in the middle of the coronavirus crisis with phishing emails. The experts warn NetWalker can break an entire computer system of a hospital.

Health professionals had to be very careful about emails. This extremely dangerous ransomware can paralyze the entire computer system of a hospital. It can be triggered by having just one email sent by a cybercriminal opened and the links in the email being clicked.

I hope you can see how threatening these hackers can be. The worst is their ability to penetrate and take over an online system very sneakily. You may see the need to safeguard your premises with security cameras and motion sensors. In the same way, you can safeguard your online systems and assets.

You want to be on your toes and secure yourself with the best online security tools and implement a foolproof ***Cyberattack Prevention Plan***. Of course, I can help you with surefire solutions that always work. We will discuss them in detail in Chapters 6 and 8.

CHAPTER FIVE

Sources of Cyberattack

Half the battle is won when we know the origins and motives of our enemy. Thus, we need to know what kind of people and groups launch cyberattacks and what programs or methods they use against individuals, businesses, and governments. You may already know cyberattacks are carried out by individuals, groups of hackers, criminal networks, and discontented employees. Let's take the time to look at them in detail.

Bot-Network Operators: These people use a comprised network of computers, called "bot-net," to remotely control the whole system and launch attacks with the intention to distribute messages that carry phishing, spam, adware, and malware. The main reason is financial gains.

Criminal Groups: Also attack networks for financial gain, yet their targets include identity theft, online fraud, and the spread of ransomware.

Hackers: Many hackers break into systems to experience excitement. They may respond to a challenge and boast of their superior skills, like LulzSec. They may be motivated by revenge, stalking, political activism, and financial gain.

Insiders: This group is particularly dangerous since they already have access to the system. They can hack for days without being noticed. They are usually discontented employees. Another dangerous insider is the poorly trained employee who compromises your online security by accident.

Governments: Espionage between governments has become a common in recent years. Cyberspace makes it an ideal place to operate. The mainly target another government's website. It is particularly dangerous when they do cyberespionage into the military operations of another government.

Phishers: This group aims to steal identities and private information like social security and credit card numbers. Their primary purpose is financial gain.

Spammers: Unsolicited emails are sent containing spyware, malware, or they simply sell various products/ services without permission.

Terrorists: The goal of this group is to destroy, weaken, or compromise important infrastructures. They want to threaten national security and disrupt the economy.

It's clear we are up against a highly trained army of cyber attackers. They are well-equipped with advanced weapons. Now, let me show you what entryways they can use to attack your personal devices, accounts, and your business.

Computer Viruses

We all know a computer virus can ruin a computer or a whole network. It's a malicious program infecting the operating system of computers or *smartphones*. It enters the system when it is attached to a common file in the form of a code. When the user executes the file, the virus is activated and becomes infectious. It then starts replicating itself in the system.

3 Types of Viruses

1. ***Viruses affecting files - called a direct-acting virus***
2. ***Viruses affecting the boot sector - called a resident virus***
3. ***Viruses which are a combination of the first two - called a multipartite virus***

The purpose can be a joke, spying of personal information, or the destruction of files and equipment. Most of the time, the process goes unnoticed to users until the device or system has been compromised.

7 Characteristics of Computer Viruses

1. They are kept dormant in the RAM of the device, even when the program is not running.

2. They infect the files on the hard drive.

3. They spread to other devices by exchanging files.

4. Since they are executable, they are a parasite housed in other programs from which they are executed.

5. They act when the programmed conditions have been met. For example, it can be a specific date, or when a certain number of replications have been achieved in the infected device.

6. They have a specific objective such as: slow down computers, destroy files, obtain secret information, etc.

7. The effects are usually persistent, even after the device is disinfected.

9 Primary Virus Types

1. *Trojans:* The name comes from the story of the Trojan Horse. They obtain personal information from the user's device without being detected.

2. *Backdoor Virus:* Similar to a Trojan, it allows a third party to introduce infected files into the system

3. *Keylogger:* It captures the keystrokes and runs without being noticed by the user. It reveals what the user is typing on the screen. It is also known as Keylogger Trojan and it is mainly used to obtain passwords.

4. *Boot Virus:* This starts when the operating system boots up.

5. *Time Bomb:* This virus stays quietly in the system until it is activated at a time programmed by its creator.

6. *Macro Virus:* Embedded in files containing macros (particularly doc, xlxs, and ppt,) when the file command is executed, the virus commands are executed first.

7. *Worms:* This virus spreads through email and replicated within a system to anyone who receives messages from an infected account.

8. *Zombie:* Allows remote control of a computer by a third party without permission.

9. *Browser Hijacker*: This virus controls the internet browser. It may change the homepage, sell advertising in pop-up windows, install tools in the bar without the user's consent, or prevent access to certain webpages.

Malware

This is also a virus that enters a system and harms or steals data. Within the malware, the virus can delete files, directories, and data without authorization. They can be spyware, worms, trojans, malicious bots which are designed to execute unwanted actions by the user; or adware invading the system with unwanted advertising.

What Activity Installs Malware?

- Open an unknown file sent by email
- Browse the internet without updated antivirus and/or anti-malware programs
- Browse unsafe networks
- Download programs/software from unknown or suspicious sources
- Open files with unknown extensions

Symptoms of Malware

- Slow processing
- Running unknown processes
- Interruption of internet connection
- Windows appear with warning messages

Spyware

This type of virus is malware characterized by tracking and collecting private information from a specific computer without the user's knowledge or authorization. As a parasitic program, it installs itself into the operating system without the user's notice. Once installed, it runs at all times when the computer is on. Once the information is collected, it is transmitted for the purpose to identify various elements they will find useful. They may want to:

- Identify browsing preferences. With this data, the computer is invaded by functions such as pop-ups. Other apps and functions may slow down the device and cause stability issues.
- Steal confidential data such as passwords and bank details.

Phishing

This is a social engineering technique that uses mass-mailing spam to collect personal data through the internet. The method fraudulently uses email to obtain data or to redirect users to a false webpage where the user is asked to submit their data.

Phishing emails usually ask the user to confirm or send certain data. They will us the excuse of technical problems, changes in security policy, or their detection of possible fraud or promotions. The email may even incorporate the coercive formula saying if the user does not take the requested action instantly, his or her account may be blocked.

The purpose is to use the collected information to make online purchases, bank transfers, or cash withdrawals, posing as the user, but without the victim's knowledge.

Pharming

Another form of phishing which involves redirecting the user from a website they trust, to another identical but false website. The user is asked to submit personal or financial details which the hacker will use to gain financially.

Spear Phishing

This very effective variant of phishing where the hacker has determined which users relate to a particular financial institution. After sending the email to these particular users the margin of fraud success increases.

Smishing

Still another variant of phishing, this involves the use of SMS on a smart phone. They use the identity of a banking institution, or trustworthy person, to capture information of a personal or financial nature

Ransomware

When malicious programs limit or block the users' access to various systems or files, unless a payment is made to the hacker(s) who created the program, it's called ransomware. Payment is made in money or virtual currency known as bitcoin. These programs have the ability to encrypt the users' files and block the contents from the rightful owner. This can disable any computer, including mobile devices such as tablets and smartphones.

This type of program targets all sizes of business, as well as private users who are connected to the internet in their homes or any public network. It can be very difficult to detect since it is hidden within a file or program, website, video, or email that has invited the user to click to open a window for its activation.

When ransomware enters a system or device, it is first downloaded, without operating immediately. Once this occurs, the ransomware delivers a warning message saying the computer has been infected. The next step is for the program to tell the cost of a key to unlock the digital device. Once the ransom is paid the key is given to free the computer from the virus.

Ransomware takes advantage of flaws in certain operating systems to attack thousands of users worldwide.

It is considered a crime, depending on its purposes. Some of these cybercrimes are created with the aim of attacking large corporations or government agencies whose documents contain confidential or valuable information.

Demands for payment may be for more than money. They can also demand payment of services such as iTunes or gift cards from sites as Amazon.

There is no guarantee the virus operator will send the key or steps to recover blocked information. Experts in cybercrimes recommend to not make the payment to rescue the data. The best way to avoid this cybercrime is to follow preventive measures beforehand.

By now, I hope you can sense the risks where you need to safeguard yourself and your business. Before we move into the chapters with even greater value against online dangers, let's briefly cover some statistics about cybercrime.

CHAPTER SIX
Cybersecurity Statistics

Three decades ago, the internet emerged as an exciting means of communication and doing business. No one could have imagined at the time; the internet would eventually rank itself among the top security concerns in our lives.

There are many factors to consider when you are protecting your organization, your information, your customers, and yourself as a user against cyberthreats. Gallup has been doing a survey of Americans since 2000 asking them to rate their worry about crime. In 2018, the **Gallup survey** included 13 crimes. Topping the list of most worrisome, 71% said they worried about having personal, credit cards, or financial information stolen by computer hackers. Second on the list with 67% was worry about identity theft.

In 2019, we witnessed massive cyberattacks on corporations and institutions. It is why many cybersecurity experts call 2019 the year of ransomware. To help you picture how serious cyberthreats are to you, I'll share a few important statistics:

Historic Data Breaches

- In 2020, a Twitter breach resulted in $121,000 in Bitcoin in 300 transactions when the breaches targeted 130 accounts of famous people like past presidents and Elon Musk
- Marriot disclosed a security breach impacting more than 5 million hotel guests in 2020
- MGM was hacked in 2019 resulting in leaked records of 142 million hotel guests
- In 2019 the healthcare industry lost about $25 billion as a result of ransomware and 93% of healthcare organizations experienced a recent data breach
- In 2019, the average cost of a data breach was $3.92 million
- The banking industry spent $18.3 million to cybercrime in 2018
- In 2018, 51% of businesses suffered a denial of service (DoS) attacks
- In 2017, one virus, called the Wannacry virus affected more than 100,000 groups in about 150 countries. About 400,000 machines were infected at a cost of about $4 billion

- The 2017 Equifax data breach affected more than 147 million consumers and cost the company more than $4 billion
- In the same year (2017) Friendfinder sites exposed the user accounts of 412 million
- In 2016, when hackers stole information of more than 57 million riders and drivers, Uber tried to pay off the hackers to delete the data and keep the breach quiet
- In 2013-2016, one of the largest breaches hacked 3 billion Yahoo accounts

Cost of Cyberattacks

- Since the pandemic started in 2020, the FBI reports a 300% increase cybercrime
- Global expenditure for cybersecurity is forecasted to exceed $133 billion in 2022
- The banking sector will lead with the largest investment in security solutions with $16 billion in 2022, compared to $10.5 billion in 2018
- Worldwide, the information security market is predicted to reach more than $170 billion by 2022

Hacking By the Numbers

- The U.S. government allocated more than $18 billion for cybersecurity for 2021
- In 2020, the average cost of a data breach was more than $3 million
- The average time to identify a breach in 2020 was 207 days
- The time between identifying a breach and full containment averaged 280 days
- In 2018, phishing and social engineering attacks victimized 62% of businesses
- Ransomware attacks on business cost an average of $133,000
- Within the first half of 2020, 36 billion records were compromised due to data breaches
- 95% of cybersecurity breaches are caused by human error
- 86% of data breaches were financially motivated
- 61% of organizations have suffered from IoT security events
- 90% of remote code execution attacks are associated with crypto-mining
- 52% of breaches were a result of hacking, 28% involved malware, 33% involved phishing, and 32% involved social engineering

- 37% of malicious email attachments are .doc and .dot and only 19.5% are .exe
- Global expenses are forecasted to be more than $133 billion in 2022
- The banking sector will invest the largest amount on security solutions, going from $10.5 billion in 2018 to $16 billion in 2022
- Cisco predicts the number of DDOS attacks worldwide will be more than 15.4 million by 2023
- According to Varonis, the average employee has access to 11 million files, and 17% of sensitive files are accessible to all employees
- More than 77% of organizations have failed to outline an incident response plan

Small Business Statistics

- 43% of cyber attacks are on small businesses
- Only 22% of small businesses take measures to encrypt their databases
- 54% if small business think they are too small to be hacked and don't have a plan for reacting to a cyberattack if it happened

Cyber Security for Individuals

- Americans lost more than $97 million to stimulus check scams in 2020
- Americans spent a record $10.8 billion online Cyber Monday 2020, according to Adobe Analytics
- In 2020, people interacted with companies online 60% of the time, compared to only 42% in 2019
- Employees were encouraged or required to work from home in 88% of organizations during the COVID crisis and many expect to continue remote work in the future
- 64% of individual Americans have never checked to see if they were affected by a data breach and 56% don't know what steps to take in a data breach event
- In 2018, Gallup reported 71% of Americans worry about having their personal, credit card, or financial information stolen by hackers, while 67% are afraid of being the victim of identity theft
- 64% of US citizens are not aware of the steps to take if the data breach occurs
- One in every 36 mobile devices have high-risk apps installed
- According to a survey by BullGuard, 76% of Americans will access public Wi-Fi and 42% admit to using their password on a public Wi-Fi connection

A Brief Look at 2020 and Beyond

In 2020 the emergence of new cybercriminal groups and targeted cyberattacks has grown. Sooner or later we all can (will) be susceptible to cyberattack. Being a victim includes large and small companies, institutions, and average citizens. It's a dire mistake to believe you don't have enough portfolio to be a target of a cyberattack. Never underestimate your value since we are all part of data, and data is what hackers are always after.

"My message to companies that think they haven't been attacked is: 'You're not looking hard enough."

~ James Snook,
Deputy Director,
UK Office for Cyber Security ~

A small business may think it doesn't have the financial potential of a large organization, which may be true. But cybercriminals know a business stoppage can spell the ruin of any small business. In addition, the small business often is a springboard the criminals can use to reach large companies if they are a supplier to them. We are all connected and so therefore all vulnerable.

Be forewarned. Be proactive. Prepare ahead of time. These are the three key ingredients in any cybersecurity plan. Cybercriminals are prepared beyond our imagination. They plan their attacks well in advance. We must be fully aware of their approach, anticipate their plan of attack, and take effective measures.

CHAPTER SEVEN
Cybersecurity Tools

"Cybersecurity tools? Humm...

I know. You are going to talk about antivirus programs.

Am I right, Josiv?"

Well, if you said it in the 1990s, you would be absolutely right.

Antivirus software immediately comes to mind when we talk about cybersecurity. Cybersecurity started with antiviruses, and they worked very well at the time. In today's era I am not talking about antivirus programs.

Security measures are advancing rapidly. So are the hackers. In chapter six we saw a few examples of data breaches, and most of them were giants in their fields such as Yahoo, Uber, Sony, and Fox. Government offices were breached like CIA.gov.

Do you think a simple antivirus program can defend you against these evil geniuses? Of course not!!!

WARNING:

Since Endpoint Protection is a highly advanced cybersecurity program, you can find a wide range of products on the market making the claim to provide Endpoint Protection.

Every program is not what it claims to be.

Antivirus programs are no longer a viable option to protect computer systems, networks, servers, or databases. In fact, hackers become so proficient hacking antivirus programs it's no more than child's play for them.

As a cybersecurity professional, I can tell you how frustrating it is when a hacker turns your security measures into the trash. Because of the growing threat, the world became desperate to find a sturdier cybersecurity solution. After the experts went through a long period of trial and error, Endpoint Security was born. Symantec released its first endpoint protection in 2007.

With the extensive range of products claiming to be endpoint protection, you need to know the differences, or hire someone like me to choose your best option. Many companies simply changed their product name from "Antivirus" to "Endpoint Protection." In reality, the programs are still simple antivirus programs which aren't effective at all. Keep reading to learn how to recognize the most reliable and effective endpoint security platform.

What is an Endpoint Device?

Any device physically an end point on a TCP/IP network. It is an internet-enabled device and includes desktops, mobile phones, laptops, tablets, servers, and even virtual environments can be an endpoint. All these endpoints become entry points for the hackers. From any of these endpoints, the hacker can move forward to access other things in the network. As you can see, securing every endpoint device in the network is essential for cybersecurity.

What is Endpoint Security?

Endpoints described in the last paragraph are the weak links in your security system. A good Endpoint Security Platform (EPP) will protect them against suspicious and hazardous activities, as well as attacks. The EPP assures all endpoint devices follow a definite level of compliance standards.

Traditional antivirus programs search for known malware codes, and they scan every file. When there is an unknown code, particularly if the code isn't attached to a file, a problem arises. This is where the EPP rises far above the protection of the traditional antivirus program. The EPP runs in the background. They have been developed to monitor behaviors instead of just searching for known signatures. These programs are proactive and can detect suspicious behavior.

Is Detecting Suspicious Behavior Enough?

Of course NOT!

We now have Next-Gen Endpoint Protection. It is a combination of detection and action. Called Endpoint Protection and Response (EDR) and is fully loaded to fight against cyber threats and maintain cybersecurity. This solution will give you peace of mind as you run your business without risks.

Differentiating EDR from Antivirus

Earlier I told you many products claim to be endpoint protection, but they aren't. Recognizing a legit Next-Gen EDR, you need to analyze the features and complete these assessments:

Does the EDR:

1. *Have an automated detection and response (ADR) system to prevent threats?*

2. *Conduct behavioral analysis to recognize malicious or irregular files?*

3. *Use Machine Learning (ML) and Artificial Intelligence (AI) to process data and determine if the file or process is a threat?*

4. *Records modifications in files and systems to protect against ransomware, which allows restoration of those files and systems to their pre-infected condition?*

5. *Has forensic capacity with real-time assessment and root causes, along with a replay of cyberattacks to assist security teams in alleviating future threats?*

6. *Endpoint detection and response which monitors devices and systems to mitigate advanced risks and quarantine infected files.*

7. *Remediates system changes automatically.*

You want to confirm all of the features listed above are part of what the provider is offering. If you can't verify each of these features, move away from the provider. The multi-level features of the EDR are what you need today.

In the professional circles of cybersecurity, it's said cybersecurity for every business should be like an onion, wrapped in multiple layers, with every layer providing a level of security. As you build up a solid business technology security plan, you protect your data from getting into the wrong hands or being compromised.

In the next chapter, we go deeper into developing a security plan with layers (like an onion,) with each layer being a piece of strong security. When we develop a security plan and it is in place, hackers have a very difficult time reaching the center core of your business. We do everything possible to never let hackers enter the center core of your data. Should it happen, the backup is still your failsafe device.

Typically, if a hacker gets though and you have to use your backup system, your cybersecurity is extremely vulnerable. It indicates you don't have a proper security system to defend against cyberattacks. You could say "it's too late" if the hacker got in and you are forced to use your backup.

We'll do a deep dive into backups in another chapter. For now, just know the importance of having a reliable and stable backup system. They can provide business continuity and instant recovery if you have a system failure or cyber compromise.

CHAPTER EIGHT

Cyber Attack Prevention Plan

As we discussed in the previous chapter, Next-Generation Endpoint Security (EDR) is the best solution today for securing your business because it helps you take proactive cybersecurity measures. Studies show 81% of all cyber breaches are in small businesses. This statistic means one in five SMBs fall victim to a cyberattack. With the implementation of EDR, these stats can go down drastically. Here's a description of time-honored ways to prevent cyberattacks.

1. Security Assessment Description

Existing vulnerabilities need to be identified. When you know your areas of vulnerability, you can protect against unauthorized access, modification and loss of information, or loss of the administration of tech infrastructure, among other risks.

The objective is to make a baseline of existing conditions. With this information, it is possible to close the current and potential vulnerabilities. Sometimes the conditions are simply lack of knowledge by the end users about business policies and best practices. It's vital all end users know this about information security, default configurations, and the existence of malicious programs within the network.

The assessment both determines the risk and classifies the level of the threat. It goes on to find proactive measures to maintain the security of the systems. First, we apply a series of checklists. From those checklists, corrective measures can be taken quickly and without considerable cost or effort. You want the assessor to quantify the impact of potential threats on the system. They will also analyze the frequency of the threats and solutions to the risk.

For greater IT security, the system will be evaluated for risks associated with the personnel in charge, external personnel, damage to the equipment, or the loss of links. They will carefully identify the absence of measures to reduce risks.

Some vulnerabilities come from software causing errors in operating systems, applications, or routines of unauthorized access. Some are hardware vulnerabilities such as maintenance failures, inadequate physical security, or a lack of protection against disasters. Likewise, data vulnerabilities such as inadequate access controls at an administrative level, the absence of a security policy, communications, inadequate control of access to the network, or personal vulnerabilities must be reviewed.

A penetration test should also be conducted to reveal risks. Called "ethical hacking" the assessor finds the areas where security can be identified and exploited. These failures and security risks may be in the technological infrastructure, at the level of services or software. With this test, it enables the business to understand the levels of risk and the impact they may face.

When a Security Assessment is Needed

It's simply a good business practice to have an expert periodically assess your systems. New vulnerabilities may occur every day and affect the technological infrastructure. In addition, these are times you want to have your security assessed.

1. ***When new servers or applications are available.*** In most cases, security is not taken into account in the development process. This can expose organizations to new risks.

2. ***When new business processes are established.*** This implies big changes in technology, people, and the tasks they carry out, so it may come with new security risks.

3. ***Regulations within an industry*** will often require periodic security assessments to remain in compliance. These are usually carried out quarterly or semi-annually.

2. Beware of Spam & Phishing Emails

In previous chapters, we talked about spam and phishing as a source of cyberattacks. This happens when you open the emails and follow the instructions. Thus, one simple way to avoid the risk is to simply not open them at all.

One of the oldest email scams is the one from a Nigerian prince. I hope you and your employees don't fall for this one now. As a precautionary measure, you should remind your staff periodically to avoid opening spam emails so you can avoid falling victim to phishing. You should also ask a professional to help you select a reliable email service with advanced features against spam so you can reduce the risk of cyberattacks.

When an Attack Happens

If you find you have been the target of a cyberattack via phishing, try minimizing the damage by accessing your account. If this is possible you may be able to prevent the hacker from exploiting your account or changing the password. You will be able to check if the hacker has modified the account settings and restore previous settings. If you aren't completely sure if the attacker has accessed your account, consider the following signs:

1. "New Login" notifications
2. Your activity log shows repeated actions like "mark messages unread"
3. Check your sent messages folder to see any messages that were not sent by you

If you pay attention, you can prevent phishing.

The reasons people will fall for these scams is because the cybercriminals play with people's interest in financial gain. They will use the promise of financial gain as bait. Their trick is to tell someone participating in a raffle will give them huge financial returns. They will ask for data such as what bank a person uses, then trick them into giving credentials.

A phishing message often has subtle errors exposing its true identity. Commonly you'll find misspellings in domain names and redirects. Always pay attention to any website asking for credentials.

1. Check the domain spelling
2. Make sure the redirected website is familiar
3. Verify the source of the information and the origin of the emails

4. Enter confidential data only on websites you know are secure

5. When in doubt, stop – don't proceed before calling the related company to confirm

3. Passwords and Limited Access Practices

One of the main sources of malware infection is the USB. All your computers may be clean and protected. But if one person uses a USB to work on an infected computer, they just have to plug in the USB to infect every computer in your system. Your computer becomes a victim of viruses without even realizing it.

This happens because of a file called "autorun.inf" with malicious code. When the USB is plugged in the file is executed automatically. It copies an executable malware to a hidden directory on the system and runs in the background. Once executed, the malware creates replicas throughout the computer system and regenerates entries in the registry. Consequently, it becomes very difficult to completely remove it once infected. The virus also configures itself to any other USB that you connect to any computer so it can continue to infect more computers.

To protect this infections, use USB port blockers so all USBs can only be executed with your permission. Some good USB port blockers are Phrozen Safe USB, USB Flash Drives Control, and Removable Access Tool.

Other security practices are:

1. Implement enhanced password policies
2. Set user screen timeouts
3. Limit user access

4. Security Awareness

To be proactive and secure you digital assets, awareness is key. Unfortunately, people don't take action until bad things happen. Very few businesses are proactive related to security risks and it is a reflection of the lack of culture of prevention throughout society.

Many organizations barely react to security issues. They do not build robust response teams or even identify required services. One of the most difficult issues to implement in any business is managing organizational change. An important habit to create is prevention amount employees. It starts with raising awareness of information security risks. I've observed those who know the importance of prevention also know the difficulties of preventing it.

Some basic security threats have been around for more than a decade. Better prevention programs must be created. It seems the new generations are born with an amazing ability to operate any device put in front of them. For this very reason, the issue of security awareness becomes more relevant. Help them create the habit of preventing security risks.

Generating and promoting prevention initiatives will reduce risks like information leakage by social engineering and other risky behaviors. Make the necessary changes so security incidents are not occurring year after year.

Prevention through security awareness is an important pillar for changing the organizational cultured regarding security issues. Each and every security awareness activity should be directed to all staff members, along with a series of recommendations for the application. When you work with awareness, it will be easier to prevent cyberattacks because your employees are part of the security team.

The efforts need to be well-directed and the communication clear. Your goal is to raise awareness about the importance of prevention and making it a habit to benefit the whole business. Information, educational resources, and awareness can be accelerated with the right tools for employees. It's a way to create a proactive culture among employees across the business and each department.

As a business owner, you are responsible for employee training. Technological toots your busines can use include training courses, knowledge bases, and collaboration tools. In addition, you will want to verify the proposed changes are actually applied and review the effectiveness of the security awareness. Your review tools may include the following:

1. *Use Strong Passwords:* Use different combinations for users and avoid writing them down anywhere. For example, you may want to resort to using easy to remember phrases. With the phrase selected, the user can use the first letter of each word and combine them with special characters to make the password stronger.

2. *Keep devices Updated:* All devices should be updated. The easiest way to make it happen is to activate the automatic updates option.

3. *Protect Digital Identity:* Provide personal information with a great deal of CAUTION. Verify the sites you visit are safe and enable privacy settings.

4. *Social Media Profiles:* Ensure profile settings are mostly private and be careful what information you post.

5. *Mobile Device Protection:* Never forget mobile devices are also targets of malware. Download application from only trusted sites.

6. *Wireless Network Security:* If not configured properly, wireless networks can be highly vulnerable. This is particularly true with public networks. Never make financial transactions on public networks.

7. *Ask For Help:* If you are concerned you may be a victim of a cyberattack or suspect a security breach, immediately contact your IT security expert.

5. Implement Advanced Endpoint Security

We described and detailed the reasons for using Advanced Endpoint Security (EDR) programs earlier in this book. We list it in this chapter because it is imperative we deal with modern challenges using the latest cybersecurity solutions. We highly recommend the implementation of these sophisticated systems in your business.

6. Multi-Factor Authentication (MFA)

First used in the 1990s, the term "Multi-Factor Authentication" or "Two-Factor Authentication" was first used to designate a hardware token generated as a One-Tim Password (OTP) related to a fixed password. As technology has evolved, with smartphones and expanded application development, gathering more factors without compromising ease of use has become common. This type of security system needs different forms of authentication to confirm validity of an activity. These are the forms commonly used:

1. Password (the user knows)
2. Security tokens (sent to the user)
3. Biometric verification (a part of the body like a fingerprint or facial recognition)

The MFA creates additional layer(s) of defense and makes it very difficult for hackers to enter the system. When only one factor is compromised, the others will continue to perform their function while the MFA makes it difficult to access the data.

The MFA system is based on two factors. However, the increase in usage of devices by consumers to make purchases and banking activities has led to the generation of 3-factor authentication for even greater security. For example, the MFA system make ask you to:

- Swipe a card and enter a PIN
- Swipe a card, scan your fingerprint, and answer a security question
- Log in to a website and enter a password sent to your mobile device or email

The purpose of an MFA is to protect your assets and accounts while blocking unauthorized access to your confidential information. Overall security of the system is improved with MFA. It offers additional protection against various types of attacks such as social engineering, remote access by Trojans, and other risks.

7. Updated Systems and Programs

Do we need to update our computers and other devices every time a new update appears?

YES WE DO!!!

Our devices make our day-to-day activities easier, but they are not perfect. We get notices from time to time about new updates for the programs installed on our devices. Sometimes it is because failures have been detected on the device or in the installed programs. It is necessary for us to install the updates form the manufacturer to be protected from the security flaws. By updating our overall system, we can avoid failures, our devices are kept safe, and our information is protected sufficiently. Browsers, software programs, plugins, and more need to be updated to keep them well protected.

It may seem a bit tedious at times, but it is a continuous and necessary activity. Because many devices and application allow us to automate the process, we may not even be aware of it happening. Don't let it overwhelm you, just make sure the updates are completed- manually or automatically.

8. What If We Don't Update Our Devices?

When a device is not updated, it leaves a big window open for hackers and cyberattacks. I can easily lead to a security breach. Our system becomes vulnerable and can be exploited by cybercriminals. They may infect our devices or carry out other malicious activities, simply from not doing an update.

The applications and programs we install on our devices are not forever. To enjoy our computers safely we need to be sure we keep them update. Similarly, features and functions are continually improved. New vulnerabilities are discovered every day allowing cybercriminals to steal information, spy on us, cause financial losses, or use our system for other criminal purposes. Fortunately, software developers and equipment manufacturers try to resolve security issues as soon as they are discovered. The solutions are offered to us in the form of updates.

Importance of Updates

Software updates are not a hassle. Any program is susceptible to security flaws. The way these flaws are corrected is with updates to the operating systems of computers, tablets, and smartphones. It is an easier and essential way to maintain the security of your systems.

We have talked about end-point security (EDR) a few times in this book already. But the tools offered by an EDR is only effective if it is updated. It's useless to have a sophisticated security program installed if it isn't capable of detecting the latest threats to your network.

NOTE: Do not confuse having the most recent update to a program with having the latest version. We can have Microsoft 2016 installed and updated, even though it is not the latest version. Manufacturers market new versions with the updates incorporated, but they also maintain the old version for a period of time by using updates.

In some cases, the security updates are called patches. You must not install anything that doesn't come from the official channels provided by the manufacturers and developers of the devices or software. Cybercriminals will use patching as a new way to infect your device. Certain websites and applications may offer you updates. If you accept them, your computer will be infected. Be careful of the source of updates.

Sometimes the cybercriminal will offer an update needing certain privileges to function properly and it can also infect your system. It is recommended you review the source of each update carefully before you proceed. Allow updates and installations from trusted sources only. Check the privileges if they are excessive or unnecessary for the intended purpose.

9. Keep an Eye on the Dark Web

The internet has its dark secrets. The most notorious is called the "Dark Web" and it is the unknown dark side of the internet. It is dangerous and contains webpages with illegal content like stolen/hacked credit cards, the sale of drugs and weapons, child pornography, terrorism, plus the confidential details of businesses and consumer data

This term should not be confused with the "Deep Web." While this web content is not indexed on search engines due to encryption, the purpose of the deep web is privacy, not illegal activity. Examples include files hosted on a cloud server, including emails.

Most web content can be divided into three categories:

- ***Clearnet:*** the part of the web containing common sites like Google, Wikipedia, Social media, e-commerce and blogs – it is also known as the "Surface Web" or the "Open Web"

- ***Deep Web:*** the parts of the web where people and organizations remain anonymous like medical records, research papers, internet banking, private networks and forums – where the data is not publicly accessible

- ***Dark Web:*** As described above, it contains anonymous networks and private communication, including illegal information

The Clearnet portion of the internets only the tip of the iceberg and it only represents 4% of the total content on the web. The vast majority of the internet is in the deep web. Being

90% of the total, it is the largest part of the iceberg. Within the deep web is the dark web and it contains 6% of the total mass of the web.

WARNING:

NEVER ENTER THE DARK WEB OUT OF CURIOUSITY!

Within the deep and dark web there are millions of pages. It is estimated between 1,000 and 4,000 million websites worldwide. The servers or networks hosting these websites are called "Darknets." The most well known of these special browsers is the TOR browser. The term stands for **The Onion Router** and it was developed with the U.S. Navy to enable browsing anonymously. It uses a series of relays within the network to assure anonymity.

The dark web is dangerous. If you want to enter the dangerous dark web, take at least twice as much caution than usual. Remember there are no protections in the dark web. With no rules to regulate it, the risk of entering this dimension of cyberspace is very high. Exposure to malware infections and personal data theft is highly likely.

On top of the risks, know these websites are watched by law enforcement. Investigators track users who access this illegal content. You may enter the dark web out of curiosity and end up being considered complicit in a crime. Hackers may find you within seconds of entering the dark web and swiftly emp all your accounts with a few keystrokes.

Dark Web = Black Market for Hackers

Common things bought and sold on the dark web include:

- Identities of all sorts of people, including C-level executives
- Small business owners who are prime targets for hackers
- Personal information such as credit cards and social security numbers
- Banking information including login credentials and date of birth
- Home and business addresses
- Credit card skimmers, fake ATM machines
- Basically anything that can be used to scam individuals and businesses

Typically this data ends up on the dark web via 3rd party websites or business who inadvertently share the data because of their lack of appropriate security measures. The list includes Target, Home Depot, Uber, Facebook, Walmart, credit reporting agencies like Equifax, and software companies like DropBox, GoogleDrive, and iCloud.

Unfortunately, most people don't even know they are breached until the hacker is able to take money out of their bank account. Even after the breach, their credentials may remain for sale on the dark web.

KRS IT Consulting knows how to enter the dark web safely. Because of this knowledge, we can monitor your work and personal email, and look for any other information capable of causing a breach. The moment we find out any information, we will alert you immediately so you can quickly change all passwords. Plus we will review any potential breach of your information with you.

10. Web Gateway Security

As we discussed earlier, cyberattacks enter your system from the internet. Web Gateway Security (WGS) can keep watch right at the entry points between the internet and your system. It is a cloud-delivered security service. This service applies constant internet security for all users, wherever they are located and whatever device they are using. This cloud-based security program spots risks on the web and through email - right at the moment they appear. The program prevents these risks from reaching your system within a few seconds. WGS avoids web intrusions and stops data from entering the system to cause infections.

11. Mobile Device Security

You can't run your business without employees. Yet, employees are the weakest link in any company's security risks. Hackers see a great and easy opportunity to steal your data and access your system through the mobile devices of your employees. If you overlook this area of security, you become a soft target for cyber attacks. Smart phones and tablets are increasing and the employee may have no perception of the security problems with the device. Still, these things can happen:

- The hacker spies on the employee's device, noticing browser use, online banking, services like PayPal. They will extract your credentials, photos, videos while they are finding a way to enter the system of your business.

- They can also take control of the device by turning it into a zombie or bot. Without you being aware, they can send spam, infect other mobile devices, or carry out any other criminal activity from your device.

- They can infect mobile devices by camouflaging viruses under the guise of programs simulating applications or games. The download will "open the doors" of their devices to cyber attacks when they thought the app was just a game. While apps for mobile devices can be found on many websites, it's recommended to download only from official stores like Google Play, Apple Store, Windows Phone, and Blackberry World. Even then, you must be careful.

- Using social engineering tricks, hackers will send malicious files through email, social media or instant messaging services. They get you to open a file to see it and tell you to download an app to enable you to see it. The file player is actually a virus you have now installed into your device.

- Hackers can take advantage of security flaws in installed applications and be able to download a virus to the device when you are downloading documents, presentations, images, videos, etc.
- Websites, sometimes legitimate ones, are manipulated by hackers. When you visit the compromised website, without the latest updated software, you are at risk of getting infected. Even memory cards and removeable accessories can transmit viruses to your device.

Implementing a Mobile Device Security program will keep you and your employees safe. When you arrange a security awareness training session for your employees, educate them about the seriousness of their mobile device security. Tell them the ways (described above) that allow hackers to get into their personal devices.

12. Firewall

You've probably heard about firewalls. It is considered to be a strong pillar against hackers. The firewall is an application serving as a barrier between your computer and computers outside your network. It can be hardware, software, or both.

Every day you connect to the internet, you are also connecting to other networks with multiple computers and users. You send and receive information as a part of regular life and business. We have already talked about viruses and hackers who may want access to your data. A firewall monitors incoming and outgoing connections and acts as a filter to traffic and block outsiders from gaining unauthorized access. It also helps to block malicious software from infecting your system.

Firewalls are your first line of security defense. They operate based on a series of rules:

1. The rule may authorize sending or receiving data. The firewall allows communication.

2. When the rule doesn't authorize the sending and receiving, it doesn't allow communication.

3. If there is no rule about it, the firewall asks the user what to do, whether to allow the action or not.

Firewalls are not the same as antivirus programs or antispyware. Yet they all usually come integrated into the same security tool. In addition, operating systems usually have a firewall. Take the time to choose the firewall you want, because multiple firewalls functioning at the same time can create issues.

Also a decision will need to be made about turning on Intrusion Detection and Intrusion Prevention and whether to send the log files to a managed Security Incident and Event Management (SIEM) system. To use the firewall fully, you need to ask for professional help.

13. Encryption

Technology allows us to encrypt information. It is the process of converting information or data into code. It is used to keep information safe from others. Encryption methods have been in use with computers since the 1970s, but the basic concept of encryption began thousands of years ago with the ancient Roman military.

When information is encrypted, it becomes unreadable. Only the authorized person can access it using a password, code, or a PIN. It is based on a complex mathematical algorithm. The processes can be carried out mainly on two content areas: storage devices and files/folders.

Storage Devices

Devices such as hard disks can be encrypted and make it impossible to access any information contained in it without knowing the encryption system. When the number of files stored on the device is large, it can take a long time to perform this task.

Files and Folders

File Level Encryption (FLE) is used by many organizations to protect their data from being accessed without authorization. This level of encryption can be on the hard disk or the cloud. While the folders and files chosen for encrypting will be unreadable for those who don't know the key, the rest of the files are available.

When we browse the internet, we may be aware of the important function of encryption to protect our activities, the information we exchange, and the data we publish. While these systems are widely used, we may not understand how they work.

Encryption must offer:

1. Confidentiality: Only authorized persons can access the information.

2. Integrity: The receiver verifies the message has not been modified during its journey.

3. Authentication: When secure communication is established between two parties, each must be able to identify the identity of the other party.

4. Non-Repudiation: The person's involved in the communication cannot successfully deny the validity of the action or information transmitted.

Encryption Types

Symmetric Encryption: When the same key is used to encrypt and decrypt the message, it is known as symmetric. Both the sender and receiver must know the key in advance. The advantage of this method is it's faster. It saves time. It is not as secure since communicating the key is highly vulnerable.

Asymmetric Encryption: Based on two keys, this type of encryption is slower, but offers better performance. You can securely communicate public keys to third parties with this method. The number of keys increases with the number of users. Key management is easier with asymmetric encryption because you only need one key pair per user to be able to encrypt messages for all other users.

Data Backup

When current data is duplicated from one medium to another the process is called a backup. It enables recovery in case of failure to the original hosting place for the data. It gives you an extra copy to use in case of an incident affecting the original data. The original data can be affected by a cyberattack, a virus from a USB, or any other cause.

At the business level, we define backup as the security of our business. It is an indispensable measure to guarantee business continuity and preserve the trust our clients have placed in our business and services. When caught without a backup, it will project a negative image and lead to mistrust. While backup is a good and important practice, I consider it the last resort in the defense line of cybersecurity.

If you handle important and confidential information, you depend on the data to keep your business moving forward. The loss of this information would mean the loss of hours of work. Serious consequences for projects can threaten the continuation of your business. As a business owner, you can't afford to miss the importance of a backup system.

Remember the place where your data is kept usually has a limited useful life. It is also subject to various risks and threats we have discussed with detail in this book. For these reasons, you must implement measures to protect the assets you store within your business. Backups are an important part of your plan.

Not all of your information assets are digital. Take the time to perform an inventory of your assets and classify them based on the critical nature of the information. Other than digital data, the information may be on paper or photographic film. The files may be text, images, multimedia, databases, programs, or applications.

Classification of Information

Consider the security measures you will apply to your information. Use this criteria:

- Level of accessibility or confidentiality
- Confidential information accessible only by management or specific personnel
- Internal information accessible only by company personnel
- Public information
- Usefulness or Function
- Customer and supplier data
- Purchase and sales information
- Personal data and internal management
- Information on warehouse orders and processes
- Impact if stolen, deleted or lost
- Damage to the company reputation
- Legal consequences
- Economic consequences
- Stoppage of business activity

Frequency of Backups

To determine how often you should create backup copies, analyze the following factors:

- Number of data files generated and/or modified
- Storage cost
- Legal obligations

Storage of Backups

You can determine the medium for data backups depending on the following factors:

- Amount of information to secure
- Cost of backup

Primary storage devices are:

Hard Drives which are easier to configure compared to other methods like magnetic tapes. They have a high tolerance/performance and storage capacity.

Cloud Storage is based on securing backups on third-party servers. Your major concern should be to confirm they have the security guarantees you need. Remember to encrypt all your backups to the cloud which will protect your data in the event of an information leak.

Advantages to cloud storage are:

- You have a backup outside of the company
- It eases business continuity issues because the data is available at any time
- When an incident occurs within the organization, the data remains protected

The downside is a concern about confidentiality. You are sending the information to an outside party. You must sign Service Level Agreements (SLA) with the provider to ensure the availability, integrity, confidentiality, and access control of the backup. The access depends on internet connection while sending or restoring backups. High upload bandwidth is required to make sure the backup copies are sent in a timely way

3-2-1 Backup Strategy

The diversification of your backups ensure you have recoverable copies. Here is the 3-2-1 rule for backups:

- First create three copies of the data – the original and two backups
- Store copies on two different storage media to protect them against different risks
- One of the copies is stored outside the business, preferably in the cloud

Risks to the Backup

To ensure backups are secured, you need a suitable place to save the copies. Over time, storage devices may deteriorate, be susceptible to mechanical failure, be subject to human error in handling (falls, contact with water,) suffer the consequences of a disaster (fire or flood.) For these reasons, keep track of the useful life of any physical backup to ensure any possible deterioration won't affect the integrity of the data.

Retention Period

The time for retaining a backup varies depending on the needs of your business and/or the legal requirements of your company. Decide how long to keep the backup copies based on:

- Is the information stored still valid or useful to your business
- The useful life of the medium storing the backups
- The requirement to keep several copies before the last one is made

Use encryption tools to guarantee the confidentiality and integrity of sensitive information you have stored. This encrypted backup is then protected in the event of theft or unauthorized access.

Schedule a restoration test periodically to ensure the restoration of the backup operates properly. Otherwise you will risk recovery. The practice will also give you a guide telling you what steps you must follow to restore the backup copy if a real contingency occurs or the regular personnel is absent. The ultimate objective of creating backup copies is having the ability to restore the data and continue business in the event of a cyber catastrophe.

Destroying Old Backups

Hackers can recover data from abandoned or broken backup media. So when you dispose of the media of backup copies, you must ensure they are destroyed safely and irreversibly. If the information reaches a third party, it could be used in a detrimental way for your company.

CHAPTER NINE
What to do Right After the Attack!

Situations may occur that seem like a cyberattack. Such things a file server being inaccessible, a website down, the inability to send or receive emails, and the like may cause concern. It may be the result of maintenance activities taking place on the system, or the system is poorly configured. Sometimes, the same type of occurrence may be due to a real cyberattack.

Not long ago many SMBs thought cyberattacks were not going to affect them. They would ask, "Why would a cybercriminal want to attack my small company when there are much bigger fish to catch?"

Earlier chapters have covered some answers to this question. Cybercriminals don't discriminate between the size of a business. Whatever size of your business - small, medium, large, high or low tech - you must be prepared to deal with a security incident.

In this chapter, you will learn what you should do if a cyberattack hits your business and what steps to follow. We will outline what order you would want to follow using those steps. Remember, no matter how many and well-implemented security measures are in your business, there will always be the risk of being attacked at any point.

Having an action plan with guidelines to follow in the event of a security and data breach needs to be done before the attack. A fundamental part of any security plan is having an appropriate response plan in place.

Determine What Happened

The first step is an initial assessment. Consider the following points to make a determination:

1. As you take these steps, record and document all the information you collect. It will be very useful for future challenges.

2. Review your business systems responsible for identifying any unauthorized access. Look at things like IDs, firewalls, logs, etc.

3. Contact the department or person in charge of managing your network to check if maintenance tasks are being carried out.

4. Identify the type of attack and its severity level. For example, a ransomware attack is different from a Denial-of-Service (DoS) attack. Other types like malware, unauthorized access, theft of information, physical damage, and others are different and will need a different approach to resolve.
5. You also want to determine if the problem is originating from an internal or external source.
6. Assess the situation at the level of its severity or criticality and use this assessment to set priority for which issue needs to be resolved.

As you take these steps, the information will be useful for preventing a future attack, whether the current situation is an attack or a false positive. If you can't determine if it is an attack, it's better for you to act as if it were an incident. When you have all the data available at the time when the attack was detected, it helps define the problem moving forward.

Identify the Scope of the Attack

An accurate assessment includes knowing what elements of infrastructure have been compromised. You will want to list all the devices, equipment, servers, networks, and other components affected. List the processes, applications, or services which depend on the affected infrastructure.

Once you know the scope, you can ask other types of related questions such as: What are the related legal or contractual implications of the incident? Do we need to report it to the official bodies regulating cybercrime, like the **FBI Internet Crime Complaint Center IC3?**

Keep a detailed record of any actions taken from the time when the incident was detected. Every precaution is important when dealing with cybercrime. Ask a series of questions like:

- What did the person who detected the incident do?
- What commands or tools were used?
- What immediate measures were taken to contain the attack?
- Did the security software alarm go off?
- Were the logs reviewed to locate other suspicious entries?

Initial Communication

A critical part the cyberattack response process is communication. Once the problem is identified, you want to know which people are aware of it. Develop a list of the immediate contacts of support staff in the event of an attack, like technical support, service providers, and forensic expert. You need this list whether the attack in internal or external.

Part of the action plan is identifying the one person responsible for coordination during the entire process. Also know which people are authorized to make decisions about the operations or services which have been compromised. Ensure each person knows how notifications related to security incidents will be communicated and how the information will be updated regarding email, telephone, or other impacted actions.

IMPORTANT: Only those people or departments who are involved in solving the issue should have knowledge of what happened. As we discussed earlier, a company's reputation is priceless. Any information leak right now could have far more negative consequences than the incident itself. Only specific personnel should be aware of the security incident. Appoint only a very responsible person to coordinate the response.

Damage Containment and Risk Minimization

The effect of a security breach can be greatly reduced by acting quickly. If a response is delayed, the situation can spiral out of control. Though each attack has a different nature, the following priorities are present in all types of attacks:

1. *Protect people's safety*. This must be the highest priority.
2. *Protect the company's valuable information*. This includes the personal information of customers, suppliers, or the business plan. When the classification of information is made, choose to protect the information marked as critical.

3. *Protect the equipment and systems.* Sometimes stopping the business processes and services is detrimental. Yet, in the long run, it is more likely to worsen the situation if you don't stop the affected systems.

Damage Containment

Consider these aspects of the business in regards to containment:

- Know the impact disconnecting all or several computers on the network, which is needed in most situations. Pay attention to the impact on any service-level agreements (SLA.)

- Determine the channel of entry used by the cyber attacker to compromise the security of your business. Take measures to protect the channel so your system can't be attacked by that route again.

- Change all access credentials for all users.

These are the first steps to demonstrate the importance of having a sturdy prevention plan in place, then implementing and monitoring the plan. To prepare for response and recovery, you also need a professional cybersecurity company working with you.

After Incident Review Documentation

As stated earlier in this chapter, documentation will help you understand the incident and evaluate the actions taken. With proper documentation at the end of the incident, you can learn from your mistakes and reinforce good practices to prevent such incidents in the future. It can also inform you about how you may want to adjust your plan based on what you learned this time.

1. Record the date and time the incident was identified.

2. Reanalyze the type of breach. Now that you have all the information about the incident, you will want to reaffirm the type of breach: ransomware, information leakage, denial of service, etc.

3. List the resources affected, along with the severity and how they affected the business. Record if it was necessary to stop services, notify customers, or suppliers.

4. Outline all the actions taken to resolve the incident along with whether they were carried out and how.

5. During the entire process, what pieces of evidence were collected? Evidence may be many things including received emails, or encrypted files or disk drives.

6. Indicate what may have been the origin of the incident based on all the steps, even if the incident has been resolved.

7. List all persons participating in one way or another while resolving the security breach.

8. Record the date and time of resolution.

As said earlier, the documentation process starts from the moment a security incident is identified. Organize the documentation chronologically and have the person in charge sign the document.

Inventory of Incurred Damage

A meticulous list covering the cost of damage caused by the breach, direct or indirect, must be compiled. You may need to provide it later as evidence in a trial for your insurance claim. These costs may include such things as:

- Reputational damage
- Replacement of equipment and software
- Legal damages
- Costs associated with recovery from normal activity
- Costs associated with the analysis of the incident
- Costs due to downtime or loss of earnings

Review and Revise Security Policies

I call this step "Lessons Learned."

When the whole process is completed and documented properly…when business operations have returned to the normal routine…take time to recap. Ask yourself the following questions:

- What failed for the incident to occur?
- What security policy didn't work?
- What needs to be improved to prevent it from happening again?
- Have I informed employees of what has happened?
- Do I need to train employees to equip them for how to act in such adverse events?
- How do we improve the steps to make response management easier?
- Was communication with the media correct when we shared news of the incident?
- Was a transparency exercise carried out with public opinion?
- Was any communication affecting private information, of customers or suppliers, carried out in a timely manner?

Restoring affected systems and services at the end of a security breach is not enough. Applying the necessary security measures to keep it from happening again is not enough either. After you have recovered daily activity in the organization, correctly documented everything that happened, assessed the damage and reviewed the company polices, you must make the final step to thoughtfully outline the "Lessons Learned." It will prepare you and the company to prevent cyberattacks in the future.

RECAP: Stages of Cyberattack Response

1. ***Identification:*** the incident is detected, the scope is determined, and a solution is formed. This stage include people who are responsible for the business, operations, communication, and contacts with technical support, forensic experts, legal advisors, and cybercrime officials if necessary.

2. ***Preparation:*** The necessary tools are gathered; such as anti-malware, file/device integrity checkers, vulnerability scanners, log analysis, backup and recovery systems, and forensic analysis.

3. ***Containment:*** Prevent the attack from spreading to other resources. Minimize the attack by separating computers from the affected network, disabling compromised accounts, changing passwords, and other similar activities.

4. ***Resolution:*** It's time to eliminate compromised elements, reinstall affected systems, and recover backups.

5. ***Review:*** Throughout the incident, all the details have been documented. The collected data is archived, and the lessons learned are discussed with cybersecurity staff. Employees are trained to prevent future risk situations.

CHAPTER TEN
Cyber-Liability Insurance Coverage

Throughout this book we have explained the extreme risk of a cyberattack. In the WannaCry incident alone, more than 200,000 computers, in all kinds of businesses, and in more than 150 countries were affected. It is but one example of how a massive attack can affect your business. The consequences of such an incident may affect commercial data, patents, finances, equipment, and - above all, the reputation of the people or the brand.

This rapidly growing risk has caused insurance companies to become involved. While we are all aware of insurance because of the different types of policies we have purchased. Cyber-liability insurance is a new field of insuring for loss. Every business should think about buying a cyber policy to protect them since cyber threats are becoming the number one business risk for organizations of all sizes. In fact, Ram Dantu, a professor at the Center for Information and Cyber Security at the University of North Texas has said, "Data is a new form of cash."

Unfortunately, until now, most SMBs have ignored the risk of cybersecurity. They often don't realize conventional insurance policies for "Material Damage and Civil Liability" does not cover this risk and most of them remain unprotected. The WannaCry incident, and all the other risks discussed in this book is enough to sound the alarm among SMBs for taking cyber-liability coverage seriously.

What Does Cyber-Liability Insurance Cover?

By now, you realize cyber risks are real. Businesses face new risks and threats every day. Cyber insurance is the best alternative to overcome a data breach. They are designed to cover the consequences of cybercrimes, especially for SMEs and the self-employed or freelancers.

Keep in mind, what is covered by this type of insurance is not yet standardized. The cybersecurity insurance market is growing rapidly. More than 100 insurers now offer digital policies and collect premiums exceeding $2 billion. **Orbis Research** projected the global market for cybersecurity will hit more than $17 billion by 2021.

Yet this is less than 1 percent of the property and casualty premiums written by insurers. Unlike most of the policies written, these insurers do not have a lot of data to support what is covered and how much to charge. The risk is rapidly growing and without historical data, it's difficult to predict the losses.

Share the steps you have taken to prevent cyber threats. The plan you developed will show you are doing all you can to avoid a claim. Also, insurers know people are the weak link in your security plan, so keep the insurance company informed about how you communicate safety measures to employees. This information could drive down the perceived risk and possibly lead to lower premiums.

The purchase of cyber risk insurance is frequent among digital businesses and it is the area where these insurance products are growing the fastest. Yet, you need to know what is covered by this type of insurance policy. The list below covers some of the things covered by cyber-liability policies, but not all policies cover all the risks I have listed here.

Company Image and Reputation

Many cyberattacks undermine the image or brand of a company. This type of damage is relatively easy for the hacker. Insurers can offer solutions to mitigate these practices. They may a variety of coverages such as legal advice, forensic computing, and public relations to manage, administer, and mitigate the incident.

Cyberattack Management

Prevention is not always enough. When a SMB or self-employed person recognizes this risk can bring the business to the verge of closure, they find a cyber insurance policy will cover various services to restore normalcy after the equipment is damaged or the data is stolen. The policy includes the expenses of computer services necessary to investigate, contain, and resolve the incident. Such things as cleaning and removal of malware, recovery of deleted or damaged data, and unlocking of hijacked computers are included.

Data Protection and Civil Liabilities

Third-party claims can arise after a data breach. As you know, data is something all businesses treasure as one of their most precious assets. It is also one of the main targets of cybercriminals. The theft of customer and supplier data can have significant consequences for a company. These third-parties consider you responsible for the protection of the data. Insurance to cover against third-party claims arising from a security breach on your network is important.

Civil liability coverage pertains to damages to third parties and employees for the theft or loss of confidential business information and personal data. It also covers the same people for the publication of any content shared as a result of the breach. In addition to all your prevention measures, having cyber liability insurance is evidence you have done everything possible to protect both the business and third-parties.

Cyber Extortion Coverage

When a cyber criminal hijacks the equipment of an SMB, it often leads to cyber extortion. They may block systems and demand a ransom in exchange for the release of data. This can put a small business at serious risk. This threat coverage includes ransoms to end extortion, and fees of specialized advisors to lead the investigation. The advisors determine the origin of the breach to end the threat or even reimbursement of the ransom money.

Compensation for Losses

Business increasingly is conducted online. Many professionals can't work without the internet, particularly freelancers or self-employed. If the network goes down, business stops and significant losses can occur. When business is interrupted or suspended due to failure in the network, causing an inability to access data, this type of coverage is responsible for the loss of income and the expenses to solve the internet blackout.

Fund Transfer Fraud

In addition to being essential to manage the company, the internet is also an important channel to make payments. Cybercriminals can make fraudulent electronic transfers at the expense of the business. This type of insurance covers the losses caused by fraudulent electronic transfers.

Misuse of Personal Data

Unintentional violation of data protection regulations and claims from third parties can become very costly. Whether the business mismanaged the protection of business data due to ignorance or lack of means, a digital security problem can be devastating. The company may even be subject to inspections and sanctions because of the improper use, control, or processing of personal data. Your insurer should cover defense expenses and any claim of administrative sanctions.

Equipment Security Monitoring Services

Insurance companies have found information security advice may not be enough, so they take a proactive role in the digital security of their policyholders. They locate cybersecurity services to install devices and applications going beyond the typical antiviruses. In that way, some insurers offer anti-hijacking applications, and surveillance and detection services to avoid the illegal use of the client's name and brand on the net.

Data Protection Legal Advice

The custody of data is a big responsibility. Companies can have online security problems and administration penalties for the improper use, control, or processing of personal data. Some insurers offer legal advice to deal with complaints and claims for their clients.

Purchasing Cyber Liability Insurance

As we said earlier, there are more than 100 companies offering cyber liability coverage worldwide. Choosing which company fits your risk can be challenging. Here are three key considerations to follow to compare your options:

1. ***Risk Identification:*** Cybersecurity involves different risk for different companies, so each liability policy will be different too. Take time to understand the cyber risks to your organization clearly to determine where you need protection most. The range of risks may be data or physical. For example data breaches or theft of corporate assets. When you have identified your risks, you can find the policy meeting your needs.

2. ***Policy Coverage:*** Cyber liability coverage can complement your existing policies. Some firms may ask for a blend of insurance products to get the most suitable coverage. It's essential you understand how each insurance product offered can help you if you're dealing with a breach.

3. ***Claim Filing Process:*** Correctly filing a claim can be tricky. Every insurer follows a procedure to confirm the authenticity of the claim, as well as the response time for issuing compensation. Many points and clauses in your policy can work against you and your claim.

You want to be clear on the process needed. You're not likely to have an insurer point out they won't pay for some particular coverages. But if you read the fine print carefully, you will find those points written in the fine print.

Ultimate Goal Of Cyber Liability Insurance

Doing business with peace of mind is of utmost importance. Cybersecurity threat is growing rapidly fueled by the growth of technology. Cyber liability coverage has become essential for business.

PART TWO

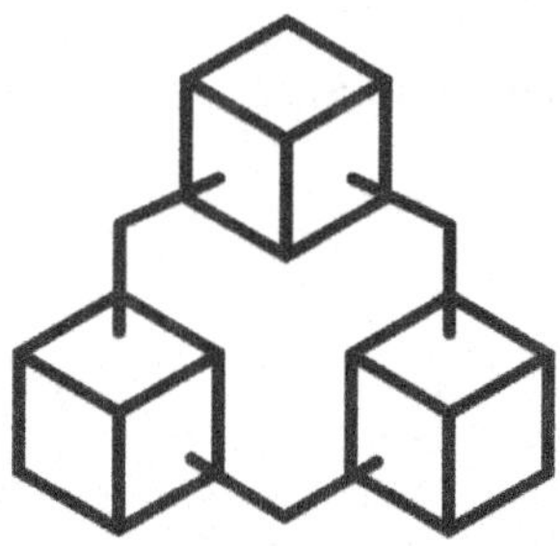

Managed IT Services

CHAPTER TEN

Managed IT Services (aka MSP)

The need for technology is increasing in the daily operations of business today. Along with increased need for technology, comes an increased in demand for IT resources. To respond to this growing need, many companies felt the need to establish their own IT department. Yet, the establishment of an IT department is not always the best solution. Startups and SMBs may recognize the importance of IT and know it can't be ignored or compromised, but it is often difficult to budget for it.

Managed IT services come into play in response to the need. The total or partial outsourcing of IT services attend to the requirements of the company on a permanent or temporary basis. This service is often referred to as a Managed Service Provider (MSP.) The MSP provides the resources a business needs. The service may be managed remotely using software allowing remote monitoring. This type of service gives the business assurance critical services of the company will be supported with the availability, capacity, performance, and security of the IT assets.

Origins of Managed IT Service

In the 1990s, a new Application Service Providers (ASPs) emerged on the scene. They offered remote support without the need to travel on-site for the customer's IT infrastructure. The duties of the ASP were outlined in a managed IT service agreement with made the service company responsible for the proper functioning of the IT services. The customer pays a monthly fee to transfer the burden of IT management and maintenance to the service provider. Monthly fees can be structured as a price per device, per user, or a flat fee.

A crucial aspect is the service level agreement (SLA) which outlines the degree of compliance in the delivery of services. The SLA benefits both the customer and the IT service provider by identifying the services to be provided and measures the degree of compliance in the delivery of services.

Remote monitoring and management of servers, equipment and mobile devices have been the most popular services. Their popularity has led to a lot of offers on the market. Service providers have developed new formulas to differentiate their offerings according to the changes in the market, especially in terms of IT security assistance.

This naturally led to service providers partnering with security providers to provide cybersecurity services. The rise of cloud computing has been a driver in the evolution of MSPs too. These are some of the most common IT service areas:

Helpdesk and Service Desk: The main function is to solve all incidents from the client or different departments within the organization.

Equipment Installation and Maintenance: The company's devices are updated and optimized for the highest performance. The same people who perform this service also install new equipment.

Logistics Management: Dedicated to the administration of technological logistics, these personnel make sure the tasks are carried out correctly.

Backoffice Management: According to the IT requirements of a business, these personnel are in charge of managing human resources, inventory, etc.

Network and Systems Management: The entire technological infrastructure of the company is administered by these personnel. They ensure all the workers can carry out their functions with maximum possible performance.

Selecting a Managed Service Provider

Contracting an MSP is an important decision for all business owners. You will want to look for these things in their profile:

- Deep Understanding of Your Business: The provider asks the right questions to demonstrate they understand your needs and strategic vision. You want them to be totally familiar with your company.

- Technology Orientation: You want to be confident the services are provided by experts in technology. This means the highly qualified personnel are continually updated with state-of-the-art technologies.

- Flexibility: Constant adaptation to your growing company is important. The provider must have the capacity to move from initial needs to what may be needed today.

Levels of IT Service

Services can be offered in a modular way where you are allowed to contract only the services you find necessary to cover current and future business needs. Examine the level of technological maturity of your company by asking these questions:

What kind of IT activity can be carried out internally with current resources?

What other activities are best outsourced to an external IT service provider?

The goal is to identify the precise level of services required.

- **Corrective Level**

The most basic form of managed IT support is activated when an incident occurs. This is a service desk support which focuses on providing a single point of contact for the comprehensive management of incidents, requests, and queries to users. The impact to business can affect one, several, or all users.

- **Preventive Level**

This administrative level provides for updated and protected infrastructure following a checklist of planned periodic tasks. The purpose is to guarantee the best availability, performance, and security by keeping information systems in perfect condition. It includes patch updates to the latest version of each manufacturer.

- **Predictive Level**

Real time analysis of the health and status information from all IT components comes from the monitoring level of support. The company gets quick response to incidents and improved management of availability, capacity, and performance. The main function of this level is focused on avoiding or mitigating incidents caused by equipment failures and intervening before they occur.

- **Strategic Level**

This level of service is like having a technology strategic partner in the goal for continuous improvement. Among other functions, the MSP helps with tactical measures to align business needs with market demands.

This is an excellent way to reduce costs and continue to innovate with new products and services for the company. Delegating the tedious tasks of implementing technology to the MSP supports the company growth and can increase the portfolio and business size to become a more competitive player in the marketplace.

CHAPTER ELEVEN
Managed Service Provider Benefits

As a managed service provider (MSP) we have seen the impact of digital transformation in companies. Including new technology in the business world is becoming increasingly important and optimizing those technological decisions leads to a clear improvement in performance and a reduction in costs. Yet, we recognize the management of all IT resources can sometimes be an overwhelming task. Having an MSP as a technology partner is a great long-term solution.

Cloud computing provides work environments which are comfortable and intuitive. Programs like Office 365 offers applications to make this new type of environment possible. Among the offerings you'll find SharePoint enables horizontal communication between multiple departments. With online versions of the classic Windows office tools, people can work from anywhere. With OneDrive, everyone is able to safely store information and files.

Manufacturers or providers make constant revisions and updates to maintain the level of these tools. Without these automated revisions, keeping up with them can generate a certain level of erosion in the technology and the IT department. When staff is trying to update manually, it often forces companies to spend money hiring professionals who know how to use the programs. A solution to this challenge is to hire an expert IT MSP.

When a company doesn't have an IT department, they will need to outsource the care of the technological infrastructure. The things outsourced may be basic installation and continued care of services such as perimeter protection in networks, installation of work programs, maintenance of tools connected to computers, and the management of updates. Having an IT provider will bring a pragmatic and effective solution to the situation.

Benefits of a Good MSP

- ***Focus on Growth and Income:*** When the maintaining of the company's technological infrastructure is outsourced, a third party can keep everything in an optimal state while the company can dedicate more time to take care of the processes directly corresponding to their growth and income.

- ***Cost Reduction:*** A technology partner is able to collaborate with the company without the costs of hiring more people to do the tasks related to IT needs.

- ***Incident Control***: Companies must adapt to the every-changing global market on a day-to-day basis. Without adapting to change problems may occur spontaneously which need an expert to deal with the aspects related to technology. The MSP anticipates small contingencies within the infrastructure. You will be able to count on the MSP to serve the departments depending on them more effectively.

- ***Cybersecurity:*** This book has outlined many of the risks to a business regarding cybersecurity. Having someone who stays aware of the risks and remedies to minimize business threats is almost a mandatory investment. The rise of cryptocurrencies, and other malicious operators planning to extort businesses with untraceable currency, is growing rapidly. Many companies will choose to not risk exposing sensitive information to attack by these cybercriminals.

- ***Reduce Staff Downtime:*** Downtime is an alarming killer of employee's productivity. You don't want to see your employees wandering around during work hours due to faults in IT operations. An IT MSP will monitor your network 24 hours every day. When an issue occurs, they are there to fix it instantly. This saves a lot of downtime for staff.

- ***Streamlined IT Budget***: Without an MSP, IT costs can be unpredictable. Systems and equipment may fail at any time and result in unforeseen expenses. When this happens, your IT budget can be ruined. When you hire and train internal staff, it is expensive and the turnover can be expensive too. MSPs usually have a fixed, monthly rate. Regular maintenance and repairs are part of the contract with them, so you don't have to bear any unexpected expenses. They also offer future upgrades too.

It is important to calculate the Return on Investment (ROI) when you plan new activities to lead to an improvement in your business. For the same reason, trusting an expert MSP will save losses and improve business productivity. This solution is proving to be one of the trends more and more firms are joining.

Questions to Ask a Prospective MSP

If you want to hire an honest, competent, responsive and fairly priced computer consultant or MSP, you must ask the right questions.

I've often asked people "What steps will you take when you need to hire a computer consultant?" These are the common steps they share in answer to the question:

- Search the internet or ask a friend for a referral.
- Make a phone call leading to finalizing the hire.
- Ask the price.
- Make a payment to hire. Voila!

These may be the USUAL steps for most people, yet a smart buyer will add an important step: Ask Questions! Getting the best answer to these questions will determine if you hire an efficient consultant or regret your decision later. These are the answers you need to get an honest, competent, responsive, and fairly priced computer consultant.

QUESTION 1: How quick and efficient are they at answering your calls? Do they answer your calls live? Do you get voice mail most of the time where they ask you to leave a message and endlessly wait for a callback? Do they answer phone calls on weekends and after-hours?

***At KRS**, we recognize we are in business because of our customers. If we can't answer you on time, there is no purpose of us being there. We answer your phone calls live from 8:00 am to 6:00 pm, Monday-Friday. For emergencies after hours (including weekends,) we have a number you can call so we can address your issues in a timely manner.*

We recognize many CEOs and executives work outside the standard working hours. When you can't access a functional computer network it is extremely frustrating. When you are working on a weekend, it's even worse. One way or another, one of our computer consultants are at your service 24 hours a day, 7 days a week.

QUESTION 2: When you call, what is their response time? Any reputable MSP can instantly tell you the metrics for their organization. If they hesitate, or fail to answer, the response time number may not be good.

At KRS, *we assign a technician to work on your issue within 15 minutes of your call, or sooner. Someone will call and resolve the issue within one hour. It's not just a verbal commitment. These response times are part of our service agreement with every client. Read the whole service agreement before you sign it. We stand behind our commitments. If it isn't part of the other MSPs service agreement, you never know what they offer you.*

QUESTION 3: Do they "geek-speak"? Some computer consultants will make you feel stupid when you ask a question. Their answer is full of "geek speak" which makes it tough to process the answers they give to you. When you spend good money hiring an expert, you don't want to deal with the technical stuff. You want someone who will spend the time to explain everything in easy-to-understand language.

At KRS, *our well-trained and experienced team members have the "heart of a teacher." Every team member answers your questions and explains everything in a language you can easily understand and process the answers.*

QUESTION 4: Do they improve your network's performance in the latest and most innovative ways? Problems will arise at some point. You don't want your consultant to wait for those problems to happen.

*At **KRS**, we believe in offering proactive solutions to our clients. We perform quarterly review meetings with each client. In this meeting, we discuss the performance overview and propose the latest solutions to improve the client's network operations. We also share ideas about reducing costs, boosting overall performance, and resolve any possible issues.*

QUESTION 5: Are invoices detailed and itemized to clearly state all the expenses? Every business network has many services and hardware involved. When the invoice is only for the overall cost of the service, it is easy to play a dirty gimmick of including "hidden charges" inside the invoice total. Some IT consultants do this on purpose.

*At **KRS**, we are transparent. Invoices are detailed and itemized for each and every service and any hardware used. The detail includes pertinent times and reasons so you can clearly see what you are paying for each month. We double-check all invoices before they are released to assure 100% accuracy.*

QUESTION 6: Do they have adequate insurance, including both worker's compensation and errors and omissions (E&O)? Unlike many service providers, the IT services industry (including MSPs) are not regulated. Inexperience or fraud can result in you losing money, time or equipment. Before we answer this question, let's talk about a true incident:

A few years back, Geek Squad (Best Buy) was slapped with multimillion-dollar lawsuits from their clients due to a technicians inappropriate behavior. The employee had accessed, copied and distributed personal information found in client's PCs and laptops which had been brought in for repairs. They even lost someone's laptop and tried to cover it up. Any computer consultant should have the proper insurance coverages.

Think about these questions:

1. Who will be responsible if a computer technician causes loss of data and your network is down for hours/days?

2. Who will pay for damages if one or more of the consultant's team members are hurt on your premises?

If your computer consultant is uninsured in either of these scenarios, can you imagine the trouble you would be facing?

At KRS*, we are as protective of you as we care for ourselves. We carry insurance policies for both E&O and worker's compensation to protect us and you, the client.*

QUESTION 7: Do they outline the timeline and budget for completing your project? Any delay can result in loss of time, money, and resources, no matter what the reason for the delay.

At KRS, *we give you both budget-friendly prices and we guarantee on-time completion of all projects. Both guarantees are given to the client in writing.*

QUESTION 8: Can you get remote monitoring of your network around the clock to keep your system updated and avoid issues like downtime, virus attacks, loss of data, etc.?

At KRS, *we watch over your computer network round the clock so we can address issues before they can turn into bigger problems. Remote monitoring is an important factor in keeping your system updated and detect issues/threats before they occur.*

QUESTION 9: What do they report and how often?

At KRS, *we prepare a weekly comprehensive report with the following information:*

- *The overall health score of your network*
- *Updates to antivirus software*
- *Security settings*
- *Patches*
- Network checks outlining the status of every machine, hard-drive space remaining, backup status, speed of the network, and performance of the network.

QUESTION 10: What information do they share with you? For example, do you have a list of software licenses, passwords, user information, hardware inventory, etc.?

You may find some computer consultants who would be reluctant to share every piece of information with you. The call it network security. However, you don't want to give up "the keys to the castle" and give up control and access. Doing so can turn out to be a threat in itself. It is your business, and your system. You are the employer. So, it's your right to know every bit of the information.

At KRS, *we keep things transparent and under your control. We share everything with properly-organized documentation. It is sent to you in electronic form and is encrypted/secure so it doesn't get into the wrong hands. We don't charge anything for this documentation. The documents are updated quarterly to ensure certain key people in your office have access to the information and know how to use it.*

QUESTION 11: Do they have a backup for their on-job technicians? Accidents and circumstances can happen to anyone, including us.

At KRS, *we have employed backup technicians to avoid any interruption in your business operations. Because we follow a detailed network documentation process and update it regularly, the backup technician can easily pick up where the regular technician left the project.*

QUESTION 12: What is meant by an all-inclusive support plan? Many computer consultants claim to offer an all-inclusive support plan. Unfortunately, some of them will never mean it. They use the fine print where there are hidden charges in their contracts. They use the fine print in their own favor. It's important you read the entire contract, particularly the fine print, before you sign the service agreement. If not, they can really surprise you later.

***At KRS**, we really mean "all-inclusive" support plan. Because every situation is different, there is no one-size-fits-all solution. We customize each agreement based on the client's needs. The contract is carefully and meticulously created based on your unique business. We want to give you everything you need for your business without any extra fees or hidden costs.*

You want to be certain what IS and ISN'T included in your computer consultant contract. This is a list of things you may find in our agreement:

- *Hardware (If hardware is included, what happens if you cancel the contract?)*
- *Software (If software is included, what happens if you cancel the contract?)*
- *Third-party software support*
- *Costs and/or consequences of early cancellation*
- *Situations of the money-back guarantee*

- *Off-site backups*
- *Restoration of your network in case of a major disaster*
- *On-site support calls*
- *Support to remote offices*
- *After-hours usage of home PCs to access the company's network*

QUESTION 13: What kind of backups are offered? Do they monitor offsite and onsite? Do they still use outdated tape backups? Yes, surprisingly, some companies use this outdated way to backup your system.

At KRS, *we don't use tape backups today because they can be extremely unreliable. We make backups every 15 minutes to let your network run with the most current backup restoration. This ensures our clients have a busines continuity product in place, and their business is guarded against any threats and can withstand any failure.*

QUESTION 14: How often do they perform periodic test restores of the backups? The frequency of these tests vary among service providers.

At KRS, *we perform a monthly "fire drill" to test restore from the backup to ensure the data can be recovered if an emergency arises.*

QUESTION 15: When carrying out any sort of project or upgrade, do they backup the network before beginning? When a project or upgrade is performed, a software glitch can occur during the process.

***At KRS,** we always create a data backup as a precaution before we do upgrades or other projects.*

QUESTION 16: Do they offer a written disaster recovery plan for your data? Without a written plan for recovery, you may not know what will happen if something adverse happens.

***At KRS,** we give a thorough disaster recovery plan to each client for their data and network. We also encourage them to do a full disaster recovery plan so their network is covered in the case of adverse events happening.*

QUESTION 17: Where is their helpdesk located? Is it outsourced?

***At KRS,** we have our in-house helpdesk located in our headquarters in New Jersey, US. Also, our customer service representatives are always friendly and helpful.*

QUESTION 18: How do the tech experts treat the time commitments and how do they dress when they come to your organization? Do they look and act professional?

At KRS, we practice absolute professional behavior in our own offices and in yours. We want you to be confident our team members will make you appreciate our presence in your location. To show our true professionalism, the experts will arrive at the committed time and will be dressed in professional attire.

QUESTION 19: Can they support your line-of-business applications?

At KRS, we will help you run your line-of-business software smoothly. It can be a bit tricky aspect of the service we provide. We will not fix faulty software. Our experts are quick to learn new technologies, so they may already have knowledge of your line-of-business application and can directly provide support. In the cases where we cannot resolve the issue you are having on our own, we communicate with the software vendor on your behalf to get it fixed.

QUESTION 20: Do they try to shift blame on you, or do they own up to their own faults? Some computer consultants are quick to tell you "That's not our problem to fix." If you don't ask this question, when something goes out of order – like the internet service, phones, printers, or other equipment – you may get surprised when these inevitable occurrences happen.

At KRS, we are always with you. Every issue pertaining to your IT related services and equipment is our issue. We own up to all our IT faults and we will resolve the issue as quickly as possible. We do everything we can to assure you have uninterrupted business operations. We find this is a unique quality you may not find in many computer MSPs.

In contrast to the quick questions asked at the beginning of this chapter, it's important you understand hiring IT services is a big decision. The company you hire will have access to the sensitive data in your organization and they are responsible for your cybersecurity. These questions are shared here with you to help you make an informed decision. Take your time to get answers in detail. Not all IT service providers are created equal and a hasty decision can be disastrous for your business.

Stay safe and well informed!

PART THREE

KRS IT Consulting

Who is KRS IT Consulting?

This entire book was written for you and I am glad you have made it to this last section. I have tried my level best to share the best of my IT knowledge and solutions in these pages. You may have learned technology is a boon and a challenge at the same time. So, you need to prepare yourself ahead of the challenges and the cyberthreats to have a smooth sail in your business.

In this section I ask you allow me to introduce my own firm, KRS IT Consulting.

KRS IT Consulting is an expert IT services and solutions provider. In a rapidly-changing and technologically-driven environment, we focus on providing cutting-edge solutions to simplify IT services and solutions. We take pride in our extensive knowledge of the industry's leading products.

Our promise is in our name. KRS stands for Knowledge, Reliability, and Simplicity. These three words represent our core business ethos.

I, Josiv Krstinovski, founded the company in 2008 and have not looked back since. We have built a reputation over time as an excellent choice for small to medium-sized businesses (SMBs.) We offer a highly personalized service for implementing wide-ranging IT functions in a way you will not find anywhere else.

Every client is assigned a dedicated IT representative to be available for mobile tech support. Even with this dedicated rep, you will most likely build a relationship with several of our IT specialists. Our goal is to remove any uncertainty or hesitancy when you have questions or problems and need answers.

What Does KRS IT Consulting Do?

Although the previous chapter talked about what is done with an MSP, let's go over a few important things you can expect from our team:

Disaster Recovery: In an uncertain business environment, filled with threats (both natural and manmade,) a lot of things can happen to your business. We ensure the utmost protection of your critical data and your business. You can have peace of mind, even if the worst should happen at any given time.

We use some of the most powerful tools for backup and restoration of your systems. One of our partners is **Datto.** Their cutting edge services allow us to assure you that when a disaster happens, you never lose the data you have built your business upon. Your server can crash and your company's data is continuously accessible. You can go back to business immediately to offer your services to clients.

Rather than get too technical, let's give you an overview of how the two primary components of the system work. This hybrid backup solution syncs between the components and gives you the confidence of knowing you are covered in ANY disaster. Each backup is scanned for any changes between the current and latest backup copy. This can be priceless in the event of a number of cybersecurity breaches or ransomware.

"We have only been with KRS IT Consulting for 1-month, and our downtime has disappeared!"

We have only been with KRS IT Consulting for 1 month, and our downtime has disappeared! Their review of our infrastructure identified numerous deficiencies that they addressed immediately. They are on top of issues before we even realize there is one!"

Laura Romanelli

Accounts Receivable at NFC

Amenity Management

Asbury Park, NJ

- A ***backup appliance*** is installed in your organization with advanced encryption technology to keep your data secure. This appliance automatically does a backup every 15 minutes. So, if your server crashes, you will immediately have access to the data less than 15 minutes before the crash. If the server needs to be replaced, you can operate on the restored data for up to 30 days while you get your insurance claim processed and the server replaced.

- ***Cloud backups*** are also created to give a copy of all your data should some natural disaster like fire, flood, or other event happens to destroy your entire physical premises, including the backup appliance and all your other hardware. Even in the worst imaginable disaster scenario, you will be able to recover quickly. Your data will be available for remote operation or the relocation of your business.

Cybersecurity: The incidents of viruses, malware, and Trojan horses continue to grow in both number and intensity. We fill the security gap and monitor your IT infrastructure 24/7. You can work with confidence at all times.

Managed IT Services: Our MSP gives a seamless IT infrastructure and it is optimized to support your business. It makes a world of difference to your bottom line. We proactively monitor and maintain your IT while ensuring you make the most out of your technology investment.

Co-Managed IT Services: Another way to improve your bottom line with the latest in technology is to involve our team with your existing IT team within your company. Your company's IT management team remains in control of the system, while certain portions of the technological needs are delegated to KRS IT. Each company requires a different structure for co-managed IT services and we adapt a strategy with you to co-create the best solution to fit your unique needs. As your organization grows, you can depend on us to step in the gaps as your partner to keep your IT running. We come to you with our expertise and save you from needing to hire, train and provide benefits like vacations, and health insurance to your growing IT team.

The services we have provided in the past are listed below. Your needs may be any combination of these, or something completely different:

- *Cybersecurity analysis and/or updates*
- *Installations and upgrades of hardware and software*
- *Creation of business analytics (which will include dashboards to keep track of your technology progress)*
- *Consultation during technology review and strategy sessions*

IT Project Management: Every business needs an IT project roadmap, developed in partnership with a team of experts who can ensure implementation is done right, on time, and within the allocated budget. We are the technology team you can count on to be there for you.

Where does KRS IT Consulting have Experience?

Our Experience: With 15 years experience, we have the ability and know-how to tackle technology's most complex problems. We've served many industries over the years. Those industries include the ones listed below and many more.

- Car dealerships
- Manufacturing and distribution
- Construction companies Real estate offices
- Medical facilities
- Corporate headquarters
- Home and home office
- Retail
- Law firms

What are the Areas of Expertise at KRS IT Consulting?

We are a one-stop-shop when it comes to IT Services and Solutions. We can help you with:

- Audio and Video Conferencing
- Data Recovery
- Desktop & Laptop Hardware Configurations-
- Disaster Recovery
- Emergency Support Services
- Mac System Engineers
- Microsoft Office 365
- Microsoft Exchange Server
- Microsoft SharePoint
- Mobile Communication Systems
- Network Cabling & Wiring
- Network Engineering
- Network Implementation & Design
- Network Security
- Remote Backup Services

"We have been with KRS IT Consulting, LLC for 5 years. They have saved us over 60% of our time and 40% of our revenue due to their outstanding services. The staff has a very knowledgeable base, and they always have a friendly and cordial demeanor. We highly recommend KRS."

Otto Acosta (IT Director - MAC Products)

- Surveillance Video
- Server Hardware Configurations
- SQL Database Administration
- Support for all Windows Server Systems
- Support for Mac OS X 10.x Systems
- Replace outdated systems like Windows 95, 98, 2000, XP, and Vista with one of the following systems: Windows XP, 7, or 10; or Windows Server 2012, 2016, or 2019
- Systems Administration
- VoIP Telephony
- VMware Server & Desktop Virtualization
- Web Development
- Wireless Networking Systems

Why Choose KRS IT Consulting?

The team members, our clients, and I have grown to become a huge family with a firm trust in one another. We believe in serving our clients with personalized support. We have a dedicated staff and in-house helpdesk so we can treat our clients like our own family. It is the reason we have grown to be one of the top-rated first. Our clients love us for our sincere and honest efforts. Below you will find a partial list of reasons to place your trust in our firm.

- *Security Focus:* While other companies will use only antivirus and firewalls, we don't think this is enough protection against today's threats. We are experts in layering innovative security pieces in your network. This guarantees your protection against ransomware and keeps your data (and your business) safe from hackers and other criminals. We take the extra step by addressing the silent threat inside your network. Employees are responsible for may cybersecurity breaches. We educate your employees on current trends and proper security protocols.

"KRS has consistently given me excellent service in the 4 years I have been with them. They make me feel very comfortable when I call, so I never have any hesitation. As a result of their work, we have increased productivity with our staff, which resulted in over 77% additional revenue that we didn't have previous years. Thank you, KRS IT for your outstanding services!"

Linda Domingues
(Rovic Transport)

- *In-House Help Desk:* We answer your calls live and connect you with an available technician within a few minutes, or we call you back within 15 minutes. This is possible because our help desk is located in our offices in New Jersey. It is how we demonstrate our dedication to providing you with the best support and services.

- ***Response Time:*** We understand having a technology challenge is stressful for everyone in your organization. So, we make it our first priority to respond and resolve your issues within an hour or less – guaranteed.

Increased Uptime: You will not pay for a month where you don't see an increase in your uptime by 20% during the month. We guarantee you will ultimately experience 99% uptime.

"KRS IT has improved our system's uptime by 99% and allowed us to be more efficient by being able to access our network remotely and securing access. Trust is important, and it's hard to find a trustworthy company that is going to provide us with great service. Thanks for all your help over the past 8 years."

– Dr. David Podell (Administrative Director – Osteo Relief Institute)

- ***Customized Roadmap:*** Our IT management program increases your productivity and employee satisfaction. We hold quarterly meetings with each of our clients to help them clearly see the results of their technology investment. In fact, our clients save money overall.

- ***No Geeky Language:*** We want you to understand everything we do for you. It's part of our devotion to customer service. We go out of our way to educate you on your technology and provide tips on how to be more efficient and protected.

- ***Hardware Provision:*** Our clients pay one flat-fee price per month and they never run on outdated technology. The use of fully upgraded hardware in technology is a serious issue for a business network. The cost of hardware equipment is expensive to replace. We will upgrade all of your equipment, including network, workstations, servers and more without an additional cost to you. In fact, you will always have a brand-new system every three years.

90-Day Money-Back Guarantee

This is the best part. We are so confident in the KRS IT services and our team members, we are willing to put our money where our mouth is. Make no mistake because you won't find many IT companies who will do this.

Yes, we offer a full 90-day money-back guarantee. If we are not doing something we promised in the IT service plan you selected and don't meet your needs, we simply give your money back and we go our separate ways.

With us, you have nothing to lose except your daily frustration and growing concerns. We are confident you will be extremely happy with our services, and team members, while being a part of the KRS family for a very long time. If you give us a try, you will see why so many businesses now can't live without our services and solutions.

THANKS FOR READING THIS BOOK

Now we want to show our appreciation with a FREE GIFT from KRS IT.

MAKE YOUR CHOICE OF 2 OPTIONS BELOW:

Gift #1

The first option is usually valued at $995. But you can get a comprehensive ***Network Health Check and Diagnosis*** as a token of our thanks.

We will review your entire network using a 27-Point Audit which will detect potential issues, security loopholes, spyware, and other hidden threats that can compromise your network and your business.

At No Charge to You, we will:

- Review system backups and restoration process when a disaster happens
- Scan your network for spyware, malware, and viruses not easily detected by most antivirus tools, let alone the removal of these threats
- Check for security updates and patches

- Review firewall and security settings
- Check the integrity of servers and hardware
- Conduct a visual scan of the server room and cabling
- Check overall performance, space, and settings for your entire system

Claim Gift #1 using this link:

https://www.krsit.com/cyberaudit/

Gift #2

If, for any reason you don't want a comprehensive ***Network Health Check and Diagnosis*** from us, you can claim 2 FREE hours of support instead. You, and lots of our competitors may think we are crazy for offering free services. We recognize some people may take advantage of us. But, since we know we are hands down the best at what we do, we won't ask you to risk your money to find out.

Here are a few things you can do with your 2 free hours:

- Discuss a project or upgrade, or even give you a free second opinion on what someone else recommended.
- Ask us to diagnose why your PCs are slow and unstable.

- Get your network's security checked for hacker attacks and viruses.
- Find out what is happening in your computer network problems.
- Analyze your network backup system to make sure it's working properly.
- Whatever IT related problem you may have, ask for two hours of FREE support as our thanks for reading this book.

Claim Gift #2 using this link:

https://www.krsitconsulting.com/activate/

Made in the USA
Middletown, DE
31 May 2023

31272305R00096